Genetic Engineering - A Glimpse of Techniques and Applications

Edited by Farrukh Jamal

Published in London, United Kingdom

IntechOpen

Supporting open minds since 2005

Genetic Engineering - A Glimpse of Techniques and Applications
http://dx.doi.org/10.5772/intechopen.77356
Edited by Farrukh Jamal

Contributors
Dr. Shaheen Shahzad, Mohammad Afzal, Shomaila Sikandar, Imran Afzal, Ana Sofia Duque, Carolina Gomes, Filipe Oliveira, Sandra Vieira, Sean Stevens, Mohammad Ashfaq, Neetu Talreja, Divya Chuahan, Werayut Srituravanich, Ranjan Singh Singh, Farrukh Jamal, Tanim Arpit Singh, Trashi Singh, Rajeeva Gaur, Prabhash Kumar Pandey

First published in London, United Kingdom, 2020 by IntechOpen
IntechOpen is the global imprint of INTECHOPEN LIMITED, registered in England and Wales, registration number: 11086078, 7th floor, 10 Lower Thames Street, London, EC3R 6AF, United Kingdom
Printed in Croatia

British Library Cataloguing-in-Publication Data
A catalogue record for this book is available from the British Library

Additional hard and PDF copies can be obtained from orders@intechopen.com

Genetic Engineering - A Glimpse of Techniques and Applications
Edited by Farrukh Jamal
p. cm.
Print ISBN 978-1-78985-179-3
Online ISBN 978-1-78985-180-9
eBook (PDF) ISBN 978-1-78985-287-5

For EU product safety concerns:
IN TECH d.o.o., Prolaz Marije Krucifikse Kozulić 3, 51000 Rijeka, Croatia, info@intechopen.com or visit our website at intechopen.com.

Meet the editor

Dr. Farrukh Jamal is Professor of Biochemistry at Dr. Rammanohar Lohia Avadh University, Ayodhya, India. His experience in the area of biochemistry includes reproductive biochemistry, plant proteins and enzymes in wastewater treatment, and insect pest management of agriculturally important food crops. Dr. Jamal started his scientific career in 1996 at the prestigious Central Drug Research Institute, Lucknow, India. He took up professional teaching in 2001 as Assistant Professor of Biochemistry at Dr. Ram Manohar Lohia Avadh University, Faizabad, India. Associated with mentoring students of biochemistry at master's level his interest in research has never ceased. At present his focus is on addressing the growing public concern of the toxicity and carcinogenicity of synthetic and recalcitrant dyes. Dr. Jamal's interests lie in designing strategies/approaches in purifying inexpensive peroxidases from easily available natural resources whose behavior under different conditions may be studied, and further protein engineering that may be done to obtain novel peroxidases that would sustain harsh conditions, culminating in overcoming the limitations in current wastewater treatment strategies. Another aspect in which he has contributed includes characterizing novel defense proteins/proteinacious protease inhibitors present in plants and their effectiveness on insect pests for applications in integrated pest management. Apart from these, he has always been fascinated by the science involved in genetic engineering and biotechnology. Dr. Jamal has undertaken several independent projects, participated in several seminars and conferences, and published his work in internationally acclaimed books and journals. He has been bestowed with a fellowship from CSIR, India. Dr. Jamal is also actively engaged in various university academic and administrative assignments.

Contents

Preface

Science has done wonders in the past decades. As the world is in the grip of Covid-19, the relevance of logical understanding is gaining prominence. One of the wonders of nature is virus and today we witness the impact of "coronavirus," which has not only created panic but has caused havoc across the world. This pandemic has brought the world to a standstill and nations are battling against it in unison. At this juncture the relevance of discoveries in biological sciences is worth imbibing. The blueprint of life is ingrained in the molecule popularly known as nucleic acid, and virus being the simplest infectious agent has the versatility to invade, conquer, and demolish the most evolved biological system on this planet. While magnification and resolution of this mysterious nucleoprotein require sophisticated equipment, xenophobia turns out to be alarming. In between the optimism to conquer and pessimism to perish is the journey of exploration with existing knowledge of science.

Altering the threads of life through genetic engineering has always been a fascinating science. The information crystallized through various research and amalgamation of technologies has given impetus to weird fantasies. In this book, the contribution by Dr. Ranjan Singh exposes readers to various strategies involved in changing the molecule of biological information. He has dwelled on the molecular scissors to molecular propagation via vectors. As one winds up seeking information about manipulating the genetic material, the chapter by Dr. Shahzad takes readers into the world of a most fascinating technology, "polymerase chain reaction," which virtually revolutionized molecular characterization, and fiddling with genetic material became much easier and less time consuming. Dr. Sean Stevens' chapter on "Genome engineering for xenotransplantation" is worthy of attention. Transplantation science has always required innovations and genome engineering could be a wonderful tool to sustain xenotransplantation.

Addressing the implications of genetic engineering, this book does not end up by limiting the technology for alteration and amplification of genetic materials from any source. Dr. Duque provides an interesting chapter on the "Prospects for the production of recombinant therapeutic proteins and peptides in plants: special focus on angiotensin I-converting enzyme inhibitory (ACEI) peptides" through genetic engineering for therapeutic purposes. Therapeutic molecules are continuously being produced and refined to work efficiently in curing various ailments. The nature of the molecule and its efficacy in the subjects being treated require sufficient attention. This chapter will expose readers to gauge the significance of therapeutic proteins and strategies to create and improve them for human needs. Lastly, Dr. Ashfaq presents a chapter on emerging technologies for agriculture using a polymeric nanocomposite-based agriculture delivery system.

Science has always been fascinating and flight of imagination has always motivated researchers to gain insight into the workings of microcosmic biological systems. This compilation follows the previous title *Genetic Engineering: An Insight into the Strategies and Application* and will turn out to be wonderful reading material for communities taking even a slight interest in biological sciences. The editor is indebted to all contributors and their efforts will further the foundation of developing nascent as well as established scientists, students, and logical minds.

Ms. Nina Kalinic Babic, the Publishing Process Manager, has always been cooperative and responsive to queries as and when required. Her contribution in shaping and finalizing the compilation for novel and ardent readers is gratefully acknowledged.

I hope that this rudimentary information will be worth reading during this Covid-19 pandemic.

Dr. Farrukh Jamal
Department of Biochemistry,
Dr. Rammanohar Lohia Avadh University,
Ayodhya, India

Chapter 1

Genetic Engineering: Altering the Threads of Life

Tanim Arpit Singh, Trashi Singh, Ranjan Singh, Rajeeva Gaur, Prabhash Kumar Pandey and Farrukh Jamal

Abstract

Over the past 30 years, the field of genetic engineering has grown in a spectacular manner. The methods involved in genetic engineering which were earlier considered cumbersome and involved sophisticated instrumentation have now became a common drill within the laboratories throughout the world. This rising technology is now involved in almost every aspect of biological research. Its application includes medical diagnosis, paternity disputes, forensic analysis, genome sequencing, etc. In the recent years, this technology has attained a large-scale attention, and now the commercial products developed using genetic engineering are known worldwide. The technique of genetic engineering is solely based on genetic information, which is encoded by the DNA in the form of genes. Through genetic engineering the genes can be introduced or manipulated within the host to develop products of value and importance, for treatment of genetic disorders, and to achieve other goals. The present chapter explains the techniques involved in genetic engineering and rDNA technology and its importance in revolutionizing different fields. The objective of this chapter is to highlight the basic principle and methodology involved in genetic engineering and its role in human welfare.

Keywords: genetic engineering, DNA, gene library, plasmid, vector

1. Introduction

Genetic engineering refers to the process of altering the genetic code of any living organisms by transferring the genes from one organism to the chromosome of another in such a way that its biosynthetic properties get modified. The manipulation of genetic material or genes is carried out using selective breeding or through molecular biological techniques [1]. This process alters the biological capabilities of an organism, and it can be utilized for the industrial production of desired proteins, enzymes, antibiotics, agricultural products, etc. [2, 3]. Genetic engineering allows us to develop crops with agronomically beneficial changes like resistance to pests and harsh environment and enhanced productivity with lesser ripening time [4]. Apart from other agricultural applications, genetic engineering can also be advantageous in curing human disease [5].

Genetic engineering can also be referred to as the mechanism of changing the level of protein expression. In a condition where a large amount of protein is required for the purpose of purification, its level of expression can be altered by changing its promoter [6, 7]. Hence, the term "genetic engineering" can also be referred to as "protein engineering" since the biochemical properties of a protein are changed through gene mutation or in vivo alteration of genes [8].

The technique of genetic engineering has evolved through our context of understanding genetics.

The methodology of manipulation of genetic material was developed in the 1970s. In vitro, the DNA was altered in a test tube and later introduced within the living cell thus altering the life process of the organism [9].

The current chapter debriefs about the outline of the genetic engineering process and its application in various fields.

2. Steps involved in genetic engineering

In a broad perspective, manipulating the DNA is done by isolating it from the cells and cleaving it using sequence-specific restriction endonuclease. Further, the two independently isolated DNA from the microbial cells are mixed and sealed using DNA ligase. Lastly, the DNA is introduced into the cells, which are grown and identified based on the altered properties of hybrid DNA [10].

For example, DNA contains a gene that is responsible for providing antibiotic resistance to the microbial cell "A," isolated and introduced into a vector (plasmid), and then transferred into bacteria "B" which gains antibiotic resistance and is a transformed bacteria.

2.1 Isolation of desired DNA fragment or gene of interest

The first important step in genetic engineering is to acquire the gene of interest which can be obtained by the methods or sources mentioned as follows:

2.1.1 Production of DNA fragments by restriction digestion

The desired DNA fragment carrying the gene of interest is cleaved from the whole DNA using restriction enzymes. These enzymes are the key and an important base of genetic engineering. There are two types of restriction enzymes known till date, i.e., exonucleases and endonuclease [11].

Exonucleases cleave the dsDNA from the terminals, whereas endonucleases cleave the dsDNA at specific nucleotide sequence present amid the center. Different varieties of endonucleases with different cleavage sites have been identified and used in the process of genetic engineering. Certain restriction enzymes like EcoRI produce single-stranded self-complementary fragment with sticky ends, whereas enzymes like Hpa I produce double-stranded noncohesive fragments [8].

Many a times there exists a certain probability that the cleavage site of restriction enzyme is available within our gene of interest, and thus the gene will not remain whole after the restriction digestion [12]. This problem can be overcome by employing hydrodynamic forces to breakdown the DNA. Sonication and homogenization are the common methods employed for the fragmentation of DNA. The DNA fragment acquired by this method is purely random, and also no sticky ends or cohesive ends are generated. Later the DNA fragments are checked for size and purity using agarose gel electrophoresis.

2.1.2 Genomic library

It comprises an entire genome of an organism that has been developed using molecular cloning methodology. The DNA of the organism is stored in population of identical vectors. In prokaryotes the genes coding for proteins are continuous, while in eukaryotes, exons (the coding region) are interrupted with introns (non-coding region). Thus, developing genomic library for eukaryotic organism remains challenging.

For construction of genomic library, the DNA from the organism is isolated and digested using restriction enzyme to get fragments of DNA of specific sizes. These fragments are later inserted into the vector using DNA ligase, and then the vector is introduced into a host organism which can be *E. coli* or yeast. The *E. coli* is a preferred host for protein production due to its rapid growth and ability to express proteins at high levels. It is also utilized in storing DNA sequences from other organisms. The genetically engineered *E. coli* is used for different studies of medical and pharmaceutical importance. Similarly, the yeast with altered genome has enhanced ability for the production of alcohol in brewery industry. The DNA fragment is later retrieved from the host cell for the purpose of analysis or study.

2.1.3 cDNA library

This library comprises mRNA purified from a cell, tissue, or entire organism which has been changed back to dsDNA using reverse transcriptase. The cDNA/complimentary DNA fragments are inserted into the host cell. A cDNA library comprises fragments of complimentary DNA which constitute certain portion of genome of the organism.

2.2 Insertion of gene into a suitable vector

A vector is a DNA molecule that has the ability to replicate inside the host to which the desired gene has integrated for cloning. The vectors include plasmids, cosmids, bacteriophages, bacterial artificial chromosome (BAC), yeast artificial chromosome (YAC), etc.

2.2.1 The vector system

They are important cloning vehicles, within which the fragment of gene of interest is inserted and transferred into a suitable host system [11]. Mostly circular DNA that has small size and is of bacterial origin is utilized as vector. The vector should possess certain characteristics to make it suitable to be used for the process of genetic engineering:

- The vector DNA should possess the ability to infect the host organism and replicate within it.
- It must possess cleavage site for specific restriction enzyme through which a foreign gene can be inserted.
- The ability of the vector DNA to replicate within the host should not be compromised after the insertion of a foreign gene.

In order to promote cloning within vector, it must possess certain important sites:

- **Ori site**—The process of replication initiates from this sequence. The presence of this site is necessary for independent replication of plasmid within the host cell. This sequence is also responsible for controlling the copy number of the gene of interest. If a large number of copies of DNA of interest are required, it must be cloned in a vector containing ori site supporting high copy number.

- **Cloning site**—It is site that is recognized by the restriction enzyme (preferable one) so that the gene of interest maybe inserted after digestion of the plasmid. The presence of more than a single cloning site may complicate the process of gene cloning as multiple fragments will be generated. The insertion of gene of interest is done at a restriction site present among two antibiotic resistance genes. For example, in the vector pBR322, a foreign piece of DNA can be joined at BamH I site of tetracycline resistance. Due to this, the recombinant might lose tetracycline resistance as its sequence is altered by insertion of foreign DNA, and thus the plasmids will lose tetracycline resistance. The transformants can still be distinguished from non-transformed ones by plating them on ampicillin medium. The transformed ones can grow on ampicillin medium but when transferred to a medium containing tetracycline, they are unable to grow, whereas the non-transformants can still grow on tetracycline medium as they still possess intact tetracycline-resistant genes.

- **Selectable marker**—Within a vector a selectable marker plays a key role in the identification of transformed cells from the non-transformed ones. Usually a selectable marker provides resistance towards antibiotics which are used for selection of transformants from non-transformants. These markers can provide resistance towards antibiotics like ampicillin, tetracycline, etc.

2.2.2 Plasmids

These are extrachromosomal DNA that possess the ability to replicate independently within the host cell. Plasmids are circular in shape and impart special properties to the organism possessing it as it may code for antibiotic resistance, bactericide production, etc. [13].

The plasmids can be classified majorly into two categories:

The first type is self-transmissible, i.e., the ones that possess the ability to promote conjugation and are transferred quickly amid the bacterial population. These kinds are not generally employed as vectors.

The second type is the non-self-transmissible, i.e., the ones that cannot regulate their own transfer among bacterial population. These categories of plasmids are extensively used in genetic engineering as vectors [14].

The gene of interest or the foreign gene is inserted within the plasmid DNA vector; for that the closed circular plasmid is cleaved using restriction enzyme to make it linear. A plasmid contains a single cleavage site for a restriction enzyme to avoid multiple digestions. The gene of interest is inserted in the linear plasmid, and then the recombinant is converted to its original circular form [15] (**Figure 1**).

2.2.3 Phage vectors

The viruses that infect the bacteria are termed as bacteriophages. They possess a very simple structure consisting of a genetic material which might be a DNA or RNA surrounded by a protein coat termed as capsid. These phages can also be employed as a vector as large piece of DNA of interest can be incorporated in the genetic material of the phage. Different phages have been developed for the purpose of genetic engineering containing a single site for the restriction enzymes [16]. These commonly used phage vectors are M13 phage, T4 phage, T7 phage, etc. The DNA of these phage vectors are altered by ligating the alien DNA fragment, and further the phage vector is inserted into the host cell [17]. These cells are later grown in a culture medium containing X-gal. The cells transformed with the recombinants will produce white colonies, and the non-transformants will develop blue colonies.

2.2.4 Cosmids

A cosmid is basically a plasmid with cos site, which contains cos sequences necessary for packaging. Cos sequences are ~200 base pair in length. Cosmids are hybrids comprising phage DNA molecule and plasmid of bacteria. Just like any other vector system, cosmids also require a selectable marker and an ori site. The cosmids lack λ-genes so unlike phages they do not produce plaques. Instead plaques colonies are formed by using selective media according to the marker just as done in the case of plasmids (**Table 1**).

2.3 Introducing vector into host cells

The membrane of the cell prevents the DNA from diffusing in or out. This barrier must be manipulated or altered so that the host cells are able to take up the foreign DNA. This goal can be achieved by different ways.

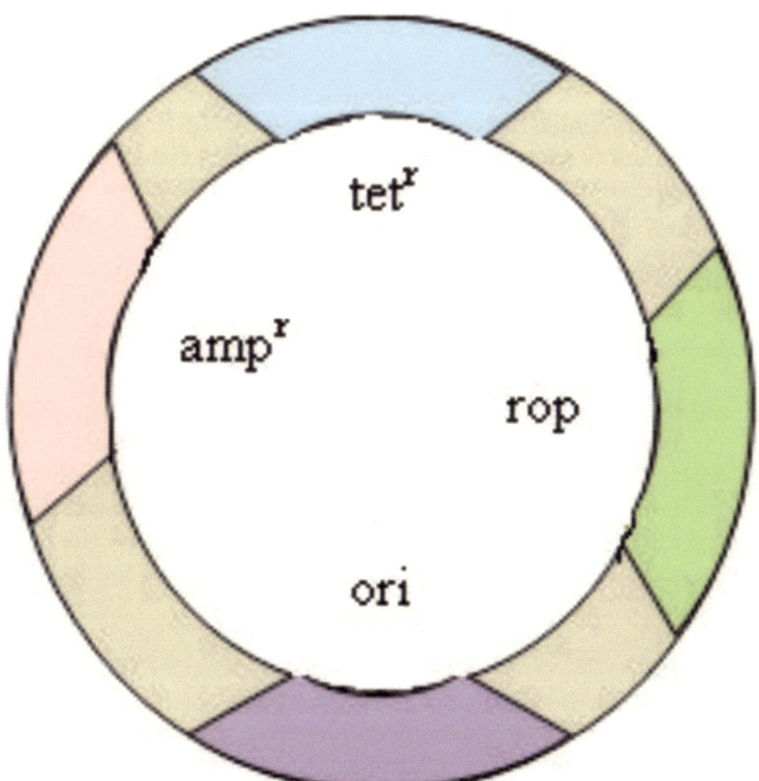

Figure 1.
Plasmid pBR322 exhibiting ori site, ampicillin- and tetracycline-resistant selectable markers.

Enzymes	Specific function
Restriction endonucleases	Cuts the DNA at specific nucleotide sequence
DNA polymerase	Addition of nucleotides at the 3′ end of the DNA
DNA ligase	Seals/joins the two fragments of DNA
Exonucleases	Remove nucleotide from the 3′ terminal of the DNA strand
Alkaline phosphatase	Removes phosphate group from both 5′ and 3′ terminals
Polynucleotide kinase	Adds phosphate group to the 5′-OH end of the DNA to promote ligation

Table 1.
Enzymes involved in genetic engineering and their function.

2.3.1 Transduction

The mechanism of inserting the foreign DNA into host cell using a phage is termed as transduction. The virus can attach themselves on the host and transfer their DNA into the cells. In genetic engineering, this ability of viruses can be utilized purposefully for the transfer of recombinant DNA into the host cells. The recombinant DNA can be transferred to a virus by inserting it within any bacteria and infecting it with the bacteriophage [18]. The virus when multiplying within the cell also intakes the recombinant DNA, and after the completion of lytic cycle, the virus containing recombinant DNA bursts out the cell. This virus now contains recombinant DNA which can be transferred to the host cell by infecting it with the phages.

2.3.2 Transformation

The mechanism by which the bacterial cell intakes DNA from the surrounding environment in which the bacteria is found or within the experimental solution is termed as transformation. This mechanism is carried out by exposing the bacteria to calcium chloride followed by heating the medium. This process will enable the bacteria to take up the recombinant DNA that we have introduced within their surroundings [15].

2.3.3 Electroporation

It is a physical method in which electric pulse is applied across the cell which creates temporary pores in the cell membrane. The DNA is then transferred using these pores within the cells. In electroporation the host cells are introduced into a conductive medium, and an electric pulse at specific voltage is passed through the solution lasting for a few microseconds. The electric current alters the phospholipid layer of the membrane, and pores are generated through which foreign DNA is transferred [14].

2.3.4 Microinjection

It is a common method that is usually employed in transferring the recombinant DNA within any plant cell. In this process the DNA is injected physically in the cells using a gene gun. The microscopic particles of tungsten or gold are coated with the recombinant DNA and are loaded to a gun. The gun comprises of high-pressure helium which ejects out the particles at a very high velocity. The outer covering of the plant cell, i.e., the cell wall, is easily penetrated using a gene gun. After entering the plant cell, the particles release out the recombinant DNA which now becomes the part of the cell [19].

3. Applications of genetic engineering

3.1 Gene therapy

Gene therapy is the process of correcting defective genes or introducing new genes into the existing cells for the cure and treatment of diseases [20]. Through gene therapy we can correct the root cause of the disease, i.e., the genes. The first approved gene therapy was employed to correct the deficiency of the enzyme adenosine deaminase (ADA) which was carried out on a 4-year-old girl Ashanti DeSilva. This girl suffered from severe combined immunodeficiency (SCID) as the gene coding for ADA was defective causing deoxyadenosine to accumulate and destroy T lymphocytes. After gene therapy, the 4-year-old developed no noteworthy side effects and grew normally into adulthood.

The gene therapy is of two types:

- In germ line therapy, the modified/therapeutic genes are transferred into the germ cells, and the individual's offspring would remain unaffected. In germ line therapy the genes are introduced into sperms and eggs.

- In somatic cell therapy, the modified/therapeutic genes are transferred into the somatic cells due to which only the treated individual will possess the modification. In this therapy the genes are introduced into the bone marrow, blood cells, etc. [6, 7].

3.2 Synthesis of insulin using genetically engineered *E. coli*

In humans, insulin is produced as prohormone and needs to be processed to work as a functional hormone. Before the advent of genetic engineering, the insulin was extracted from the slaughtered cattle. This insulin when injected to the human patients in majority of cases induced allergies. The active human insulin comprises chains A and B linked together by disulfide bridges. The synthesis of insulin using genetic engineering was achieved by Eli Lilly, an American company in 1983 [5]. They isolated the human gene responsible for synthesis of insulin and introduced it into *E. coli* through a vector. The chains from the host cell were extracted, separated, and joined through disulfide bridges.

3.3 In agriculture

The genetic engineering has revolutionized the agriculture sector and resolved the problem of feeding the ever-increasing population to a certain bit. Using genetic engineering different varieties of plants have been developed that possess better agronomic characteristics.

The crops developed through genetic engineering are termed as transgenic or genetically modified crops. The transgenic crops can be produced by transferring the gene of interest within the plant to obtain the desirable traits. The characteristics of plant can also be altered by silencing or removing their own genes.

The important characteristics of transgenic crops include:

- Crops with better ability to tolerate environmental conditions like drought, cold, salinity, etc. For example, the salt-tolerant transgenic maize plant was produced by incorporation of gut D gene from *Escherichia coli* [4]. Crops resistant against insects, herbicides, and other chemicals. For example, Sulphonylurea-resistant tobacco plants are produced by incorporating the mutant acetolactate synthase (ALS) gene from *Arabidopsis* [4].

- Biofortification—The improved crop variety has enhanced nutritional qualities as compared to that of the conventional crops. For example, the golden rice is very rich in vitamin A.

- An altered rate of growth for higher productivity in less time, e.g., genetically modified tomatoes, potatoes, tobacco, etc. [21].

3.4 Production of insect resistant Bt cotton using genetic engineering

The bacteria *Bacillus thuringiensis* synthesize certain proteins with the ability to kill insects like dipterans and lepidopterans. This toxin was termed Bt toxin which remains inactivate, but when engulfed it gets activated due to the alkaline pH of the gut. The insects engulfing it eventually die. Through genetic engineering the genes responsible for production of Bt toxin were isolated and were used in crops like cotton to make it pest resistant. The toxin responsible for killing the pest was coded using *cry* gene. The Bt cotton was resistant against budworms, beetles, mosquitoes, and flies [22].

3.5 Transgenic animals

The animals whose genetic makeup has been altered to express foreign genes are known as transgenic animals [23]. The transgenic animals are important as they enable us to:

- Study the regulation and expression of gene.

- Study genes responsible for causing disease in the humans. Transgenic models allow us to infect them with diseases like cancer, CFTR, etc. and to develop an effective treatment against it [22].

- Ensure safety of treatment that the human receive. Transgenic mice are developed to study the effect of vaccines before using it on humans [24].

- Produce biologically important products like human proteins, protein-enriched milk for human babies, etc. [25].

4. Conclusions

Genetic engineering has become an integral part of the modern era, and its vast applications have touched and evolved almost all important fields necessary for living. This technology has revolutionized medical, agricultural, and pharmaceutical sectors. The genetic engineering involves a group of techniques that are employed for varied purposes based on our requirements. This technology has provided us the cure for many diseases, pest-resistant crops, transgenic organisms and hormones, and other important products. The modern era of biological research remains incomplete without this technology. Genetic engineering holds the future of modern medicine, agriculture, and pharmaceutical industries. The genome editing tools are gaining much attention in the present era. Genetic engineering is the answer to all the solutions that we are facing today. In order to feed the rising population of the world, it is the tool for enhancing crop productivity. Through genetic engineering the fermentation ability of yeast can be enhanced for increased alcohol production. Similarly all sectors like antibiotic production can also be augmented, and the structure can also be modified for better efficacy. Genetic engineering also opens the gateways for the treatment of diseases that have medically no potential cure available till date. The technique of gene manipulation can provide better longevity of life. This technique is the future of almost all sectors, and with the advent of time it may provide solution to all the present problems.

Author details

Tanim Arpit Singh[1], Trashi Singh[2], Ranjan Singh[3*], Rajeeva Gaur[4], Prabhash Kumar Pandey[5] and Farrukh Jamal[6]

1 Department of Biosciences, Maharaja Ranjit Singh College of Professional Sciences, Indore, MP, India

2 Department of Biological Sciences, R.D. University, Jabalpur, MP, India

3 Department of Biotechnology, Choithram College, Indore, MP, India

4 Department of Microbiology, Dr. R.M.L. Avadh University, Ayodhya, UP, India

5 Department of Biochemistry, University of Allahabad, Prayagraj, UP, India

6 Department of Biochemistry, Dr. R.M.L. Avadh University, Ayodhya, UP, India

*Address all correspondence to: ranjan.singh13@gmail.com

References

[1] Cre´pineau F, Roscoe T, Kaas R, Kloareg B, Boyen C. Characterisation of complementary DNAs from the expressed sequence tag analysis of life cycle stages of *Laminaria digitata* (Phaeophyceae). Plant Molecular Biology. 2000;**43**:503-513

[2] Adebami GE, Adebayo-Tayo BC. Development of cellulolytic strain by genetic engineering approach for enhanced cellular production. In: Kuila A, Sharma V, editors. Genetic and Metabolic Engineering for Improved Biofuel Production from Lignocelluloses Biomass. Netherlands: Elsevier Science Publishing Co. Inc.; 2020. pp. 103-136

[3] Cohen PE, Wertz DC, Nippert I, Wolff G. Genetic counseling practices in Germany: A comparison between East German and West German geneticists. Journal of Genetic Counseling. 1997;**6**:61-80

[4] Jhansi Rani S, Usha R. Transgenic plants: Types, benefits, public concerns and future. Journal of Pharmacy Research. 2013;**6**:879-883

[5] Burgess-Brown NA, Sharma S, Sobott F, Loenarz C, Oppermann U, Gileadi O. Codon optimization can improve expression of human genes in *Escherichia coli*: A multi-gene study. Protein Expression and Purification. 2008;**59**:94-102

[6] Yang T, Guo Y, Gao N, Li X, Zhao J. Modification of a cellulose system by engineering *Penicillium oxalicum* to produce cellulose nanocrystal. Carbohydrate Polymers. 2020;**234**:115862

[7] Yang G, Mahadik B, Choi JY, Fisher JP. Vascularization in tissue engineering: Fundamentals and state of art. Journal of Biomedical Engineering. 2020;**2**:012002

[8] Arber W. DNA modification and restriction. Progress in Nucleic Acid Research and Molecular Biology. 1974;**14**:1-37

[9] Clarke A, Parsons E, Williams A. Outcomes and process in genetic counseling. Clinical Genetics. 1996;**50**:462-469

[10] Sinsheimer RL. Recombinant DNA. Annual Review of Biochemistry. 1977;**46**:415-438

[11] Hedgpeth J, Goodman HM, Boyer HW. DNA nucleotide sequence restricted by the R1 endonuclease. Proceedings of the National Academy of Sciences of the United States of America. 1972;**69**:3448-3452

[12] Nathans D, Smith HO. Restriction endonucleases in the analysis and restructuring of DNA molecules. Annual Review of Biochemistry. 1975;**44**:273-293

[13] Yew NS, Wysokenski DM, Wang KX, Ziegler RJ, Marshall J, McNeilly D, et al. Optimization of plasmid vectors for high-level expression in lung epithelial cells. Human Gene Therapy. 1997;**8**:575-584

[14] Archibald JM, Rogers MB, Toop M, Ishida K, Keeling PJ. Lateral gene transfer and the evolution of plastid-targeted proteins in the secondary plastid-containing alga *Bigelowiella natans*. Proceedings of the National Academy of Sciences of the United States of America. 2003;**100**:7678-7683

[15] Zhang Y, Buchholz F, Muyrers JPP, Stewart AF. A new logic for DNA engineering using recombination in *Escherichia coli*. Nature Genetics. 1998;**20**:123-128

[16] Jones KW, Murray K. A procedure for detection of heterologous DNA

sequences in lambdoid phage by in situ hybridization. Journal of Molecular Biology. 1975;**96**:455-460

[17] Li Z, Karakousis G, Chiu SK, Reddy G, Radding CM. The beta protein of phage lambda promotes strand exchange. Journal of Molecular Biology. 1998;**276**:733-744

[18] Morita S, Kojima T, Kitamura T. Plat-E: An efficient and stable system for transient packaging of retroviruses. Gene Therapy. 2000;**7**:1063-1066

[19] Kurtzman AM, Cheney DP. Direct gene transfer and transient expression in a marine red alga using the biolistic method. Journal of Phycology. 1991;**27**:42

[20] Recillas-Targa F, Valadez-Graham V, Farre CM. Prospects and implications of using chromatin insulators in gene therapy and transgenesis. BioEssays. 2004;**26**:796-807

[21] MacKenzie D. How the humble potato could feed the world. New Scientist. 2008;**26672**:30-33

[22] Andrews LB, Fullarton JE, Holtzman NA, Motulsky AG, editors. Assessing Genetic Risks. Implications for Health and Social Policy. Washington, DC: National Academy Press; 1994

[23] Rudolph NS. Biopharmaceutical production in transgenic livestock. Trends in Biotechnology. 1999;**17**:367-374

[24] Pursel VG, Pinkert CA, Miller KF, Bolt DJ, Campbell RG, Palmiter RD, et al. Genetic engineering of livestock. Science. 1989;**244**:1281-1288

[25] Yom HC, Bremel RD. Genetic engineering of milk composition: Modification of milk components in lactating transgenic animals. The American Journal of Clinical Nutrition. 1993;**58**:299-306

Chapter 2

Polymerase Chain Reaction

Shaheen Shahzad, Mohammad Afzal, Shomaila Sikandar and Imran Afzal

Abstract

Polymerase chain reaction (PCR) is an efficient and one of the most common methods used in biological sciences for in vitro multiplication of a target DNA molecule. The technique has significantly contributed in changing and developing different fields of biological sciences since 1980s. PCR has a vital role in supporting the processes involved in genetic engineering, particularly the cloning of DNA fragments used to modify the genomes of microorganisms, animals, and plants. Consequently, the technique has numerous applications in fundamental and applied research in medicine agriculture, environment, and bio-industry. The main focus of this chapter is to describe briefly the principles, methodology, various types, and applications of PCR in different fields. Besides, different components of PCR, trouble shooting during the execution, and limitations of the techniques are also outlined.

Keywords: PCR, primer, DNA template, nucleotides, sequence, polymerase

1. Introduction

Polymerase chain reaction (PCR) is one of the most commonly used method in modern molecular biology. The technique was developed by the Nobel laureate, Kary Mullis, in 1984. It is an in vitro process to multiply a target molecule of DNA with extreme precision, making it easy to be handled and examined by routine molecular biological methods [1–4]. Since its inception, PCR has significantly contributed in changing and developing biological sciences. The first PCR machine was introduced in market in 1988. The Human Genome Project has been result of PCR based approaches [5, 6]. Owing to its wide range of applications, numerous variants of PCR techniques have emerged over the past few decades [2–4].

PCR begins with the separation (denaturation) of the strands of a target DNA molecule (known as template) followed by annealing (hybridization) of oligonucleotide primers to the target template. The annealed primers provide a start for the DNA polymerase point to add new nucleotides (deoxynucleoside triphosphates or dNTPs). The sequence of nucleotides to be added is determined by the template. This entire process of the amplification of template, i.e., separation, annealing, and polymerization, is accomplished in vitro by cyclical alterations of temperature [2, 4, 7–10]. DNA polymerases used in PCR originate in thermophilic microorganisms, largely archaea, thriving temperature between 41 and 122°C. This ability to withstand high temperature is requited in PCR to melt or separate the double-stranded DNA. Today, PCR has become a main stay in biotechnology, genomics, diagnostics, systematics, and many more areas [2–6].

2. Principles of PCR

PCR typically involves a series of 20–40 repeated temperature changes, called cycles, with each cycle usually comprising three discrete temperature shifts (**Figure 3**) [2, 4, 7–10].

1. Denaturation: 94–96°C

2. Primer annealing (depending on the primer): 45–60°C

3. Primer extension: usually 72°C

The cycling steps often start and end with a temperature step called "hold" where product extension is performed at (>90) and (~72°C), respectively. The final product is kept at 4°C before its analysis or storage. The most of PCR methods typically amplify DNA fragments of up to ~10 kilo base pairs (kb). However, some techniques can amplify up to 40 kb.

3. The components of PCR

Setting up a basic PCR requires many ingredients, reagents, and conditions which are described below (**Figure 1**) [2, 4, 7–10].

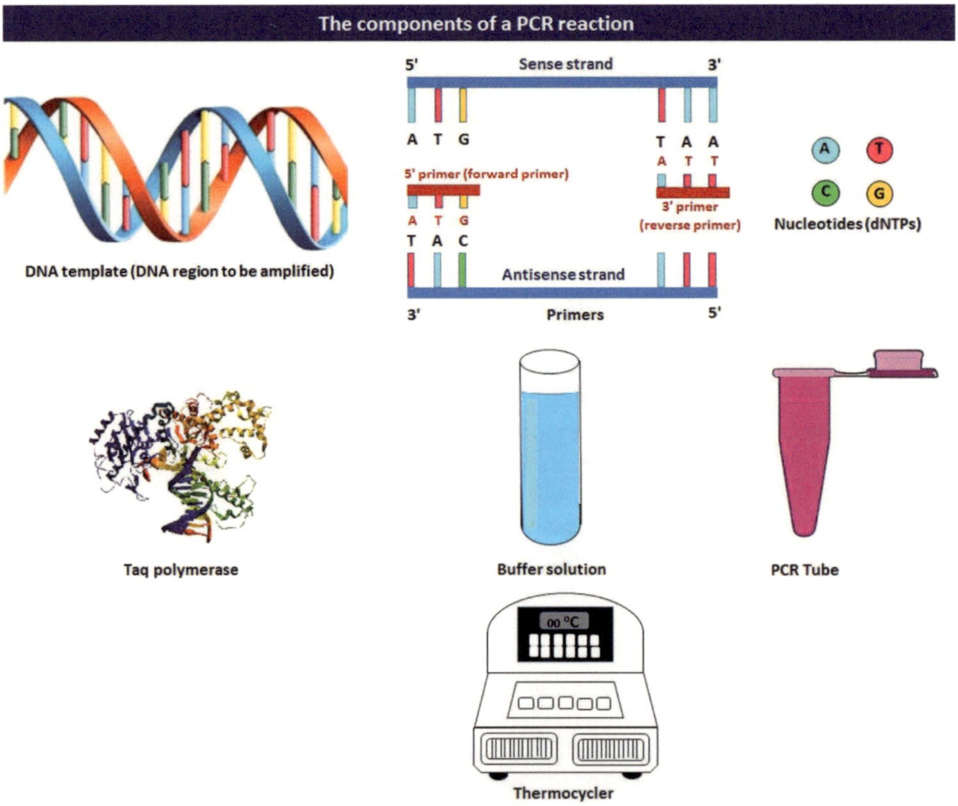

Figure 1.
The component of PCR.

3.1 DNA template

The double-stranded DNA molecule amplified by PCR is called the target or template DNA. The template defines the sequence in which new nucleotides are added during the PCR process [11]. This process is carried out in vitro by cyclically varying temperature, enabling separation of DNA strands, hybridization of primers, and polymerization.

DNA isolated from any source can be used as a template for PCR provided that it contains the target sequence. The DNA used in PCR can be isolated from blood, tissue, forensics specimens, paleontological samples, or microbial/tissue cells grown in the lab. Whatever the source, we need to have some information of the target DNA sequence, so that primers for PCR can be designed [2, 12]. The PCR primers can be designed very easily nowadays owing to the plethora of software tools that only requires target sequence information. If the sequence information is not known, the designing of primers becomes very challenging. This problem can be circumvented by using degenerate primers [2].

3.2 Primers

Primers are single-stranded DNA molecules usually synthesized commercially, i.e., polynucleotides of variable sizes [2, 7–10]. These short polynucleotide DNA strands have a free 3′ hydroxyl group, also called as 3′ end. The free 3′ hydroxyl group on the primer is needed by the DNA polymerase to add new nucleotides during the polymerization process, thereby synthesizing a new complementary strand [12, 13]. The binding of DNA primer to the target requires the separation of two complementary DNA strands (Denaturation) which is generally achieved by heating process. To perform PCR, two primers are needed to enhance both the strands of the template: a primer for one strand (or sense strand), called the "forward primer," which is the beginning of the template, and another primer for the complementary strand (or the antisense strand) called the "reverse primer." Thus, both the primers bind to 5′ ends of the sense and antisense strand.

The length of primers plays an important role in correctly identifying their designated target complementary regions. Increasing the length of primers improves their chances of matching the target (specificity). These primers are usually commercially synthesized with their size ranging between 18 and 25 nucleotides.

Primers should bind (anneal or hybridize) to the template with good specificity and strength to ensure amplification of the correct sequence. The specific temperature that is needed for primer annealing also depends on the primer sequences, e.g., the longer the primer, the higher the annealing temperature. Therefore, the maximum specificity and efficiency of PCR depends on optimal primer sequences and appropriate primer concentrations [2, 7, 8]. This in return depends on the way primers are designed and used. Improper primers may amplify undesired DNA segments (non-specific products), lower the yield of specific products, or completely fail the results of PCR. These undesired outcomes can be circumvented by designing and validating primers that preferentially bind to their target sequences. The online IDT Sci Tools Software Oligo Analyzer 3.1 and Primer Quest are invaluable aids both in primer design and validation [14]. These software tools also ensure that the two primers do not contain sequences that are complementary to each other. If primers contain self-complementary sequences, then hybridization will occur to each other, and they form "primer-dimmers." Consequently, the primers will fail to bind to their target template, leading to a compromised PCR efficiency. In addition, presence of complementary sequences within a primer leads to the formation of hairpin loop structures [15].

As the bonding of guanine and cytosine bases (GC) is stronger than that between adenine and thymine (AT) bases, primers having GC at 3′ end should be preferred for a strong bonding with the template [16]. However, the primers should not contain runs of three or more C or G bases, as this may lead to nonspecific binding to G- or C-rich sequences (mispriming) in the DNA which is not the target sequence [17].

3.3 DNA polymerase

Discovery of DNA polymerase in 1955 was the onset of PCR technology, which exploits the ability of bacterial DNA polymerase to make a complementary strand of a target DNA [2, 4, 7, 8, 18]. DNA polymerase starts making a new DNA from the 3′ end of the template. The 3′ end of the two template stands is where the primers bind which are then extended by the DNA polymerase. The most commonly used DNA polymerase is Taq DNA polymerase isolated from *Thermus aquaticus*, a thermophilic bacterium. Taq polymerase extends the DNA chain by adding ~1.0 kb per min with the enzymatic half-life achieved at 95°C in 40 minutes. Alternatively, the DNA polymerase from *Pyrococcus furiosus*, called Pfu, is also used widely due to its 3′–5′ exonuclease activity (proofreading) which is not present in Taq DNA polymerase. Proofreading allows Pfu to remove incorrectly added nucleotide during polymerization and therefore to synthesize new DNA with minimum errors. A recombinant DNA polymerase, KOD DNA polymerase, derived from the thermophilic solfatara bacterium *Thermococcus kodakarensis* KOD1 type strain, functions optimally at 85°C with 3′–5′ exonuclease proofreading activity, resulting in blunt-ended DNA products [19, 20]. KOD DNA polymerase exhibits high fidelity and processivity for small amplicons. However, for the amplicons over 5 kb, the amplification is lowered due to strong 3′–5′ exonuclease activity of the enzyme [5]. This problem can be solved by mixing wild type with the mutant form of the enzyme (with lower 3′–5′ exonuclease activity), which can result in more correct amplification of the amplicons between 5 and 15 kb [21]. Other sources of DNA polymerases used in PCR include thermophilic species like *Thermus thermophilus* (Tth) and *Thermus flavus* (Tfl) [18].

3.4 Nucleotides

PCR requires four different deoxynucleoside triphosphates or dNTPs to synthesize new DNA strands: adenine(A), guanine(G), cytosine(C), thymine(T). The dNTPs are usually provided at a concentration of 200 μM in the reaction mixture [22]. The concentration of these four dNTPs must be equal in the reaction mixture, as unequal concentration of even a single dNTPs leads to misincorporation of nucleotides by the DNA polymerase.

3.5 Buffer solution

The function of PCR buffer solution is to provide suitable conditions and chemicals to the DNA polymerase for optimal activity and stability [23]. The buffers often contain Tris-Hcl, KCl, and sometimes $MgCl_2$. PCR buffers are often available in 10× concentration and are sometimes Taq formulation-specific including the compounds shown in **Table 1**.

3.6 Monovalent cations

Potassium chloride (KCl) is normally used in a PCR amplification of DNA fragments at a final concentration of 50 mM [24].

Component	Function
100 mM Tris-HCl (pH 8.8 at 25 °C)	Maintains reaction pH
500 mM KCl	Stabilizes primer-template annealing
15 mM $MgCl_2$	Cofactor for DNA polymerase
0.8% (v/v) Nonidet P40 (Optional)	Suppresses secondary structure formation

(Thermo Fisher Scientific™B16: https://www.thermofisher.com/order/catalog/product/B16)

Table 1.
Concentrations of PCR buffers.

3.7 Divalent cations

Magnesium ions are needed by the DNA polymerase enzyme as a cofactor. The divalent cations may include magnesium or manganese ions; generally, Mg2+ is used, but Mn2+ can be utilized for PCR-mediated DNA mutagenesis, as higher Mn2+ concentration increases the error rate during DNA synthesis [25].

3.8 PCR tube

PCR is performed in a small, thin-walled plastic tube called PCR tube. The tube is specifically designed to permit favorable thermal conductivity equilibration during thermal cycling.

3.9 Thermal cycler

A thermal cycler or thermocycler is a device used to rapidly heat and cool the reaction mixtures and cycle them between the three PCR temperature steps [26]. Many modern thermocyclers employ the Peltier effect to achieve this temperature ramping, which is done by reversing the electric current [11]. Modern thermocyclers are also provided with heated lids to prevent condensation of reaction mixture during PCR operation. Older thermocyclers lacked this feature, and the evaporation was prevented by applying oil or wax balls on the surface of PCR mixture.

4. Procedure

Each cycle or round of PCR comprises three major steps, viz., denaturation, annealing, and extension, repeated for 30 or 40 cycles on a thermocycler (**Figure 2**) [2, 4, 7–10]. A number of parameters determine the range of temperature and the duration of each cycle step (**Figure 3**), e.g., the polymerase used for DNA synthesis; melting temperature (Tm) of the primers; and the concentration of reagents used, i.e., divalent ions and dNTPs. The melting temperature depends on the length and specific nucleotide sequence of a primer. At Tm, half of the DNA molecules are in the single-stranded form.

4.1 Denaturation

It is the first cycling step that involves heating the reaction mixture to 94–98°C for 20–30 seconds. Such higher temperature disrupts the hydrogen bonding of the two complementary strands to produce the single-stranded DNA templates. Thus, denaturation prepares the DNA template for the binding of primers.

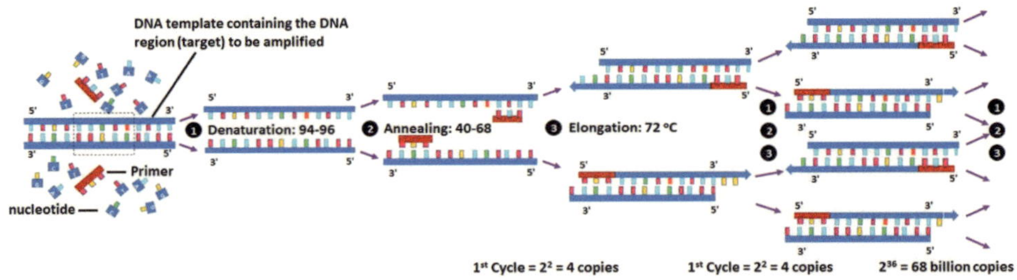

Figure 2.
A basic PCR protocol—DNA synthesis cycle.

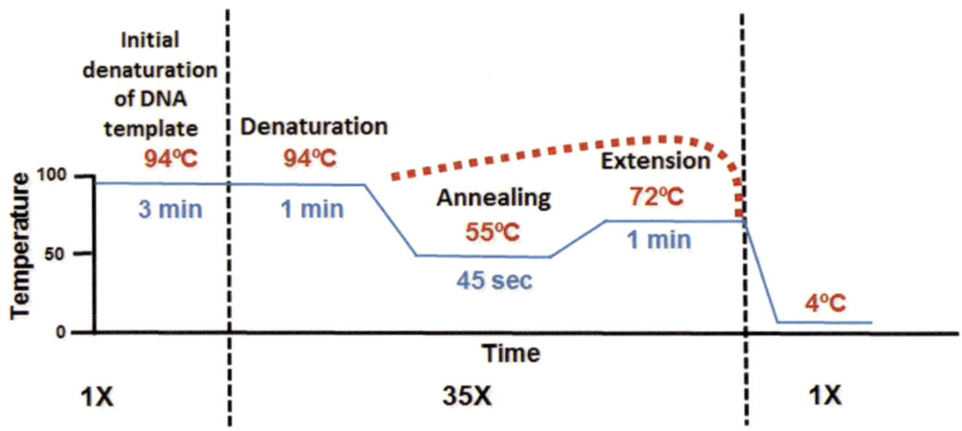

Figure 3.
PCR—a simple thermocycling protocol.

4.2 Annealing

After denaturation, the next cycling step is annealing, in which the temperature of the PCR reaction is decreased to 50–65°C and kept for 20–60 seconds. This promotes hybridization between primers and single-stranded templates. Optimal annealing occurs at temperatures that are 3–5°C less than the primer Tm. The primers should have sufficient length and GC content to strongly bind to their target template during annealing.

4.3 Extension

DNA polymerase synthesizes (polymerizes) new DNA molecule by adding dNTPs complementary to the template bases in a 5′–3′ direction. The temperature and extension time depend on the type of DNA polymerase used: Taq polymerase performs optimally between 75 and 80°C. However, the enzyme is routinely used at 72°C. The extension time also depends on the length of the template.

A single cycle of PCR comprises the entire processes of denaturation, annealing, and extension/elongation. Under ideal conditions (optimal temperature ramping, presence of substrates/reagents, absence of inhibitors), the quantity of target DNA is doubled at the end of each cycle, resulting in exponential amplification of the specific DNA segment.

Final elongation: This single step is optional and performed at 70–74°C. The final elongation may take for 5–15 minutes after the last PCR cycle and allows any remaining single-stranded DNA to be fully extended.

Final hold: The final step may be employed for short-term storage of the reaction by cooling the reaction chamber to 4–15°C for an indefinite time.

5. Magnitude of amplification

Multiple cycling steps of a PCR can exponentially increase the copies of target DNA template to millions. The number of DNA copies produced by a PCR can be calculated using the formula "2^n", where n is the number of cycles [27, 28]. For example, a PCR set for 36 cycles results in 68 billion copies of the template. Under optimal conditions, even with minimal efficiency, a PCR in 50 ml volume may produce 0.2 mg of 150 bp DNA from 100 template molecules after 35–40 cycles, with the molar weight of the fragment equal to 99,000 Da.

6. Validating PCR

The degree of a PCR success can be determined in many ways [3, 29–36]:

1. Ethidium bromide (EtBr) can be used for the staining of amplified DNA product [31, 32]. It has UV absorbance maxima at 300 and 360 nm, and an emission maximum at 590 nm, and being a DNA intercalator, EtBr inserts itself between the base pairs in the double helix. The detection limit of DNA bound to ethidium bromide is 0.5–5.0 ng/band.

2. A three primer combination approach can provide a more cost-effective end-labeling of PCR products: (i) fluorescently labeled universal primer, (ii) modified locus-specific primers, and (iii) 5′ universal primer sequence tails [34].

3. Agarose gel electrophoresis: The most commonly employed validating method, gel electrophoresis, makes use of electric current to separate charged molecules like DNA using gel as molecular sieve. Gelling agents can be agarose (for DNA >500 bp) or polyacrylamide (<500 bp). Different DNA sequences are separated out based on their sizes. DNA staining dyes (like EtBr) are applied to the gel to help visualize the DNA bands using UV transilluminator [34, 36]. The presence of a correct size DNA band (confirmed using a DNA ladder) indicates that the target sequence was present and that the PCR has amplified a correct product. Absence of any DNA band indicates that the target DNA was absent, while the presence of incorrect size DNA band indicates production of a spurious product [37].

4. The direct sequencing is often not practiced due to in accessibility or cost of DNA sequencer or even the time needed to undertake such an analysis. However, restriction enzyme digestion can also be used to assess the sequence of an amplicon indirectly [29].

7. Types of PCR

Owing to ever-growing applications, a wide variety of PCR techniques have emerged over the past few decades [2–4]. Some of the variants are mere optimization close to the basic PCR to fulfill the specific needs. Others have undergone massive modifications to suit novel applications in different biological, biomedical, agricultural, and environmental fields [6].

7.1 Conventional PCR

This is a standard PCR in which a single-primer pair is used to bind to the two separated target strands. The primers also define the target sequences that will be copied. The PCR generates millions of copies of the target DNA sequences [2–4].

7.2 Multiplex-PCR

It is a special type of PCR for the detection of pathogenic microorganisms by using several pairs of primers annealing to different target sequences in a single sample [2–4, 38]. The multiplex-PCR is mainly used to identify exonic or intronic sequences to detect mutations, deletions, insertions, and rearrangements in pathogenic specimens.

7.3 Nested-PCR

It is used to increase the specificity of DNA amplification by reducing the nonspecific amplification [2–4, 39]. The two sets of primer pairs are used for a single locus point in two successive PCR reactions. The first round of PCR is performed with a primer pair that anneals to the sequence that flanks the target region. This generates a much larger DNA product that includes the target sequence. The second PCR is performed with a primer pair that precisely anneals to the target sequence, internal to the product of first PCR. This ensures that only the correct product is amplified in the second PCR [7]. Although Nested PCR improves specificity of amplification, it has disadvantage like primer-dimer formations [40].

7.4 Real-time PCR/quantitative PCR (qPCR)

A qPCR is a technique used to quantify the amplification of a template DNA in real time during the PCR reaction. This type of PCR is commonly employed to estimate the number of DNA targets present in a sample or to study and compare the gene expression [7, 37]. When real-time PCR is used quantitatively (qPCR), the amount of amplification is measured either by using a nonspecific fluorescent dyes or sequence-specific DNA oligonucleotide fluorescent probes [4, 41]. When quantitative PCR is used above/below a certain amount of DNA molecules, it is called semi quantitative real-time PCR. Although the quantitative real-time PCR has many applications, it is more frequently used in basic research and diagnostic purposes. There is a growing industrial use of the technique, e.g., quantification of microbial load in processed foods, detection of GMOs, quantification of pathogenic viruses, etc. [42–44].

7.5 Hot start/cold finish PCR

This technique reduces nonspecific amplification during the initial stages of a PCR [4, 7, 45]. To prevent nonspecific amplification at lower temperatures, hybrid polymerases are used which remain inactive at ambient temperature and is only activated at higher temperatures. Inhibition of the polymerase activity at ambient temperature is done by using an antibody or covalently bound inhibitors. Simply, in this technique the reaction components are heated to the DNA melting temperature (e.g., 95°C) before adding the polymerase.

7.6 Touchdown PCR (step-down PCR)

This type of PCR is designed to minimize nonspecific amplification by gradually decreasing the primer annealing temperature in the successive cycles. PCR is started with initial cycles having an annealing temperature 3–5°C higher than the primer Tm. The annealing temperature is then gradually decreased to 3–5°C lower below the Tm. The higher annealing temperature increases the specificity of the primers at initial stages of the reaction, while the lower temperature permits more efficient amplification later at the end [4, 7, 46].

7.7 Assembly PCR or polymerase cycling assembly (PCA)

This technique is used for the synthesis of long DNA molecules from long oligonucleotides with short overlapping segments, alternating between sense and antisense directions. The process begins with an initial PCR with primers that have an overlap, followed by a second PCR using the products of the first PCR as the template to generate the final full-length DNA structure [4, 7, 47].

7.8 Colony PCR

It is a convenient high-throughput technique used to confirm the addition of DNA insert in the recombinant clones and their uptake by the bacterial cell. A single set of insert specific primers are designed for the areas of the vector flanking the site where target DNA fragments are already inserted. This results in the amplification of the inserted sequences. The technique is used for the screening of bacterial colonies transformed with the recombinant vectors and to perform PCR without initially extracting the bacterial genomic DNA [3, 4, 7, 48].

7.9 Methylation-specific PCR (MSP)

It is a variant of PCR used to identify promoter hyper-methylation at CpG islands in cell lines and clinical samples, including fresh/frozen tissues. The target DNA is first treated with sodium bisulfite, which transforms the unmethylated cytosine bases in to uracil, which pair with adenosine of the PCR primers. The modified DNA is then amplified using two types of primers that only differ at their CpG islands. One primer set anneals to DNA with cytosine (corresponding to methylated cytosine), while the other anneals to DNA with uracil (corresponding to unmethylated cytosine). The MSP technique provides quantitative information about the methylation when used in quantitative PCR [3, 4, 7, 49, 50].

7.10 Inverse PCR

This type of PCR is used to detect the sequences that surround the target DNA (flanking sequences). It involves a series of restriction enzyme digestions and self-ligation. The primers amplify sequences at either end of the target by extending outward from the known DNA segment [4, 7, 51].

7.11 Reverse transcription-PCR (RTP)

In this technique, the PCR is preceded by a reaction converting RNA into cDNA using viral reverse transcriptase. The resulting cDNA is used as a template for a second conventional PCR. The technique is widely used in the detection of RNA viruses and to study gene expression [7, 52, 53]. A variant of the RTP, called differential-display reverse transcription-PCR or RNA arbitrarily primed PCR (RAP-PCR), is used to study and compare the gene expression of organism grown under different conditions. The variant employs the use of short and random 10-mer or 11-mer radio-labeled primers that are annealed at low stringency conditions to promote the extension of random sequences during the first PCR cycle. This is followed by high-stringency cycles to extend the products of first cycle. The resulting products are analyzed using standard sequencing gels, and RAP-PCR fingerprints are visualized by autoradiography. The technique is extremely useful in studying tissue-specific and condition-specific gene expressions [54, 55].

8. Variants of PCR

In addition to the above mentioned techniques, numerous other variants of PCR are in use to serve a wide variety of research, diagnostic, and industrial needs, e.g., after exponential PCR, allele specific PCR, asymmetric PCR, arbitrary PCR, core sample PCR, degenerate PCR, dial-out PCR, digital PCR, high-fidelity PCR, hot start PCR, in silico PCR, inter-sequence PCR, ligation-mediated PCR, mini primer PCR, nanoparticle-PCR, overlap-extension PCR, solid-phase PCR, splicing by overlap/overhang extension PCR, suicide PCR, thermal asymmetric interlaced PCR, etc. Some of the important variants of PCR are described below:

8.1 Extreme PCR

In extreme PCR the concentration of primers and polymerase is increased 10–20 times; the amplification rate of instrument reaches about 0.4–2.0 s/. When the primers' concentration is more than 10 mol/L, the polymerase concentration is 1 mol/L, and the extreme PCR is suitable for rapid detection of virulent infectious and bioterrorism pathogens [56].

8.2 Photonic PCR

It is achieved by fast heating and based on energy conversion, thus shortening the PCR time. The specific process is carried out by using electronic resonance light emitting diode. The energy conversion process is more rapid than the conventional cooling process, causing amplification of target DNA within 5 min and thus making the PCR detection more convenient and fast [57, 58].

8.3 COLD-PCR

It is a low denatured temperature-PCR for enriching mutant genes by reducing the reactive temperature of PCR. The basic principle is founded on the base mismatch in any strand of DNA affecting the denaturation temperature. Therefore, the denaturation temperature of the mutant DNA is often lower than that of wild type DNA. The assay is often used for viral gene mutation [59] detection, cancer associated gene mutations (p53) [60, 61], EGFR, KRAS, etc.), beta globulin (HBB) mutations that cause beta thalassemia [62], etc.

8.4 Nanoparticle-PCR

Gold nanoparticles have superior electrical, optical, thermal, and catalytic activities and have the same properties as single-stranded binding proteins (ssb), which bind to single-stranded DNA and do not interact with double-stranded DNA. Therefore, the amplification effect of high GC template can be significantly improved by adding gold nanoparticles as additives to slowdown or touchdown PCR reaction systems [63].

8.5 HPE-PCR

It is an amplification technique for templates with long DNA chains and large numbers of CTG repeats involving the increase in the denaturation temperature of PCR to solve the problem of high content of DNA (G+C) [64].

8.6 LATE-PCR

It generates high concentrations of single-stranded DNA that can be analyzed at the end point using probes which hybridize over a wide temperature range [65].

8.7 Digital PCR

Digital PCR (dPCR) enables precise and sensitive quantification of nucleic acids in a wide range of applications in both healthcare and environmental analysis. It is based on detection in two discrete optical channels, focused on the quantification of one or two targets within a single reaction [66]. The technique has become a promising quantification strategy that combines absolute quantification with high sensitivity.

9. Applications of PCR

The PCR technique and its several advanced variants act as powerful tools with specialized applications which were once impossible by the scientific world [67, 68]. This versatile technique brought enormous benefits and scientific developments such as genome sequencing, gene expressions in recombinant systems, and the study of molecular genetic analysis, including the rapid determination of both paternity and the diagnosis of infectious disease [69, 70]. It enables the in vitro synthesis of nucleic acids through which a DNA segment can be specifically replicated in a semiconservative way. It generally exhibits excellent detection limits [71, 72]. It has significantly transformed

Application	Description	Reference
Diagnosis of infections	PCR approaches are used to specifically and sensitively diagnose infections (bacterial, viral, protozoan, fungal). They are routinely used in clinical laboratories to confirm and quantify these infectious agents	[71, 74]
Diagnosis of genetic defects	PCR-based detection systems are used to accurately detect (before disease onset) and confirm (after the onset) many genetic disorders	[74–76]
Diagnosis and prognosis of cancers	PCR-based approaches can identify cancer genes and analyze their expression to determine genetic predisposition to certain cancers, confirmation of cancer type, their prognosis, and treatment	[74, 77]
Phylogenetics	Phylogenetic analysis of organisms routinely relies on PCR amplification of phylogenetic markers to identify and classify them	[78, 79]
Archeology	Ancient DNA (aDNA) recovered from archeological remains are usually degraded and are in low amounts. Such miniscule quantities of aDNA are amplified using PCR techniques to improve their quality and quantity to make them analyzable for archeological study	[80–82]
Recombinant DNA technology	PCR techniques are used to generate hybrid DNA with ease and precision. The techniques are also employed to clone DNA in to specific vectors to get protein expression	[83, 84]

Application	Description	Reference
Metagenomics	Gene-targeted metagenomics combines PCR with metagenomics to identify rarest members of a sampled community and rare genes in the community members	[85, 86]
Site-directed mutagenesis	PCR-based approaches are commonly used to insert mutations (deletions, additions, and substitutions) at specific locations in a gene to study role of specific amino acids in the structure and function of proteins	[87, 88]
Personalized medicine	PCR technologies are employed in pharmacogenomics and pharmacogenetics to track genetic markers that determine the response of individuals to treatments and are used to design tailor-made drugs and to prescribe drugs in effective doses	[74, 89]
Forensics sciences	The power of PCR is employed to amplify poor quality and quantity DNA samples from crime scenes and make them reliably analyzable.	[67, 68, 90]
DNA profiling	DNA profiling methods utilize PCR-based approaches to exploit the polymorphic nature of DNA (SNPs, DNA repeats, etc.) to study the structure and diversity ecological communities, phylogeny, and population genetics	[90]
Gene expression profiling	Reverse-transcriptase PCR and qPCR are routinely employed to profile the expression of genes and to validate transcriptome profiles generated through techniques like microarray and RNA-seq	[91–93]
Identifying medicinal plants	PCR-based DNA barcoding is a tool that utilizes specific DNA sequences to rapidly and accurately identify medicinal plants species from other morphologically similar plants. This approach is also used by ecologists and conservation biologists to identifying endangered and new species	[94]
Detecting GMO	PCR techniques are used to quickly and reliably track the presence of genetically modified organism in food and feed to ensure their regulation and protection of consumer rights	[95, 96]
Meat traceability	PCR methods are used to identifying and quantifying adulteration of meat in raw and processed food.	[97–99]

Table 2.
The most important applications of PCR.

the scientific research and diagnostic medicine. Over the years, it has become a vital of clinical and diagnostic research. It has a wide range of applications in almost every field of science, for example, clinicians widely use the technique for disease diagnosis. Biologists, including agriculturists, clone and sequence genes using PCR and rapidly carry out sophisticated quantitative and genomic studies. Now for criminal identification, PCR assays are commonly employed. DNA fingerprinting is also used in paternity testing, where the DNA from an individual is matched with that of his possible children, siblings, or parents [67, 68]. Besides, PCR has enormous role in diagnosing genetic disease, whether inherited genetic changes or as a result of spontaneous genetic mutations, is becoming more common. Diseases can be diagnosed even before birth. Even PCR can also be employed with significant precision to predict cure of diseases [73]. The most important applications of PCR are summarized in **Table 2**.

10. Limitations of PCR

Since it discovery in 1980s, the PCR technique has brought about significant changes in biological sciences. Huge scientific undertakings like the Human Genome Project have been possible due to PCR-based approaches [5, 6]. It is a very sensitive and flexible technique to amplify DNA of interest. A very small amount of the target DNA can be used as a starting material. Even old or degraded DNA samples may yield successful amplification. However, there is also a long list of PCR limitations. High-quality DNA amplification needs information about target DNA sequence. The sensitivity of PCR is also its major disadvantage since the very end result of a PCR is highly susceptible to contamination or false amplification. Therefore, amplification of DNA by PCR may not be 100% specific. Moreover, the specificity of amplification is dependent on physicochemical parameter, such as temperature and Mg^{++} concentration. The PCR is also inhibited by the presence of certain chemicals such as ethanol, phenol, isopropanol, detergent compounds like sodium dodecyl sulfate (SDS), high salt concentration, chelators, etc. There is an upper limit to the size of DNA that can be synthesized by PCR. Additionally, analysis and product detection usually take much longer time than the PCR reaction itself.

Author details

Shaheen Shahzad[1*], Mohammad Afzal[2], Shomaila Sikandar[3] and Imran Afzal[3]

1 Genomics Research Lab, Department of Bioinformatics and Biotechnology, Faculty of Basic and Applied Sciences, International Islamic University, Islamabad, Pakistan

2 Consultant for Health Research and M&E, Pakistan

3 Department of Biology, Lahore Garrison University, Lahore, Pakistan

*Address all correspondence to: drshaheen@iiu.edu.pk

References

[1] Sárosi I, Gerald E, Girish VN. Polymerase chain reaction and its application. Orvosi Hetilap. 1992;**133**:16-21

[2] Bermingham N, Luettich K. Polymerase chain reaction and its applications. Current Diagnostic Pathology. 2003;**9**(3):159-164

[3] Rajalakshmi S. Different types of PCR techniques and its applications. International Journal of Pharmaceutical, Chemical and Biological Sciences. 2017;**7**(3):285-292

[4] Patel SV, Bosamia TC, Bhalani HN, Singh P, Kumar A. Polymerase chain reaction (PCR). Agrbios. 2015;**13**(9):10-12

[5] Mullis KB. The unusual origin of the polymerase chain reaction. Scientific American. 1990;**4**:56-65

[6] Garibyan L, Avashia N. Research techniques made simple: Polymerase chain reaction (PCR). The Journal of Investigative Dermatology. 2013;**133**(3):e6. DOI: 10.1038/jid.2013.1

[7] Zádor E. The polymerase chain reaction. Theoretical course: Basic biochemical methods and ischemic heart models. Supported by: HURO/0901/069/2.3.1 HU-RO-DOCS. 2011. Available from: http://www.academia.edu/35361310/The_polymerase_chain_reaction_Theoretical_course_Basic_biochemical_methods_and_ischemic_heart_models_Supported_by_HURO_0901_069_2.3.1_HU-RO-DOCS

[8] Dubey VK. Proteomics and Genomics. Lecture 37: Polymerase Chain Reaction. Available from: http://nptel.ac.in/courses/102103017/module37/lec37_slide1.htm [Accessed: 24 May 2018]

[9] Saiki RK, Scharf S, Faloona F, Mullis K, Hoorn GT, Arnheim N. Polymerase chain reaction. Science. 1985;**230**:1350-1354

[10] Joshi M, Deshpande JD. Polymerase chain reaction: Methods, principles and application. International Journal of Biomedical Research. 2010;**2**(1):81-97

[11] Bell LE. Cooling, heating, generating power, and recovering waste heat with thermoelectric systems. Science. 12 Sep 2008;**321**(5895):1457-1461

[12] Ye J, Coulouris G, Zaretskaya I, Cutcutache I, Rozen S, Madden TL. Primer-BLAST: A tool to design target-specific primers for polymerase chain reaction. BMC Bioinformatics. 2012;**13**(1):134

[13] Olerup O, Zetterquist H. HLA-DR typing by PCR amplification with sequence-specific primers (PCR-SSP) in 2 hours: An alternative to serological DR typing in clinical practice including donor-recipient matching in cadaveric transplantation. HLA. 1992;**39**(5):225-235

[14] Owczarzy R, Tataurov AV, Wu Y, Manthey JA, McQuisten KA, Almabrazi HG, et al. IDT SciTools: A suite for analysis and design of nucleic acid oligomers. Nucleic Acids Research. 2008;**36**(Suppl_2):W163-W169

[15] Vallone PM, Butler JM. Auto dimer: A screening tool for primer-dimer and hairpin structures. BioTechniques. 2004;**37**(2):226-231

[16] Li LC, Dahiya R. MethPrimer: Designing primers for methylation PCRs. Bioinformatics. 1 Nov 2002;**18**(11):1427-1431

[17] Lexow P, Ragnhildstveit E, LingVitae AS. Method for characterising polynucleotides. United States Patent Application US 10/553,113 [Accessed: 02 April 2009]

[18] Cline J, Braman JC, Hogrefe HH. PCR fidelity of Pfu DNA polymerase and other thermostable DNA polymerases. Nucleic Acids Research. 1 Sep 1996;**24**(18):3546-3551

[19] Benson LM, Null AP, Muddiman DC. Advantages of *Thermococcus kodakaraenis* (KOD) DNA polymerase for PCR-mass spectrometry based analyses. Journal of the American Society for Mass Spectrometry. 1 Jun 2003;**14**(6):601-604

[20] Morikawa M, Izawa Y, Rashid N, Hoaki T, Imanaka T. Purification and characterization of a thermostable thiol protease from a newly isolated hyperthermophilic *Pyrococcus* sp. Applied and Environmental Microbiology. 1 Dec 1994;**60**(12):4559-4566

[21] Nishioka M, Mizuguchi H, Fujiwara S, Komatsubara S, Kitabayashi M, Uemura H, et al. Long and accurate PCR with a mixture of KOD DNA polymerase and its exonuclease deficient mutant enzyme. Journal of Biotechnology. 15 Jun 2001;**88**(2):141-149

[22] Lorenz TC. Polymerase chain reaction: Basic protocol plus troubleshooting and optimization strategies. Journal of Visualized Experiments. 2012;**63**:3998. DOI: 10.3791/3998

[23] Hwang IT, Kim YJ, Kim SH, Kwak CI, Gu YY, Chun JY. Annealing control primer system for improving specificity of PCR amplification. BioTechniques. 2003;**35**(6):1180-1191

[24] Huggett J, O'Grady J, editors. Molecular Diagnostics: Current Research and Applications. Caister: Academic Press; 2014

[25] Innis M, Gelfand D. Optimization of PCR: Conversations between Michael and David. In: PCR Applications. 1999. pp. 3-22

[26] Lund K, Rothenberg BE. Applied Chemical, Engineering Systems Inc. Thermal Cycler. United States patent US 6,558,947 [Accessed: 06 May 2003]

[27] Vos P, Hogers R, Bleeker M, Reijans M, Lee TV, Hornes M, et al. AFLP: A new technique for DNA fingerprinting. Nucleic Acids Research. 1995;**23**(21):4407-4414

[28] Barnes WM. PCR amplification of up to 35-kb DNA with high fidelity and high yield from lambda bacteriophage templates. Proceedings of the National Academy of Sciences. 15 Mar 1994;**91**(6):2216-2220

[29] Telenti A, Marchesi F, Balz M, Bally F, Böttger EC, Bodmer T. Rapid identification of mycobacteria to the species level by polymerase chain reaction and restriction enzyme analysis. Journal of Clinical Microbiology. 1993;**31**(2):175-178

[30] Millipore Sigma. Technical Guide to PCR Technologies—Assay Optimization and Validation. Darmstadt, Germany: Merck KGaA; 2018. Available from: https://www.sigmaaldrich.com/technical-documents/articles/biology/assay-optimization-and-validation.html [Accessed: 30 May 2018]

[31] Mariani BD, Martin DS, Chen AF, Yagi H, Lin SS, Taun RS. PCR—Molecular diagnostic technology for monitoring chronic osteomyelitis. Journal of Experimental Orthopaedics. 2014;**19**

[32] Yap EP, McGee JO. Non isotopic SSCP detection in PCR products by ethidium bromide staining. Trends in Genetics. 1992;**8**(2):49

[33] Nakayama H, Arakaki A, Maruyama K, Takeyama H, Matsunaga T. Single-nucleotide polymorphism analysis using fluorescence resonance energy transfer between DNA-labeling fluorophore, fluorescein isothiocyanate,

and DNA intercalator, POPO-3, on bacterial magnetic particles. Biotechnology and Bioengineering. 2003;**84**(1):96-102

[34] Blacket MJ, Robin C, Good RT, Lee SF, Miller AD. Universal primers for fluorescent labelling of PCR fragments—An efficient and cost-effective approach to genotyping by fluorescence. Molecular Ecology Resources. 2012;**12**(3):456-463

[35] Gariyan L, Avashia N. Research Techniques made simple-PCR. The Journal of Investigative Dermatology. 2013;**133**(3):6

[36] Querci M, Jermini M, van den Eede G, editors. Training Course on the Analysis of Food Samples for the Presence of Genetically Modified Organisms. JRC-European Commission, World Health Organization—Regional Office for Europe; 2006

[37] Huang T, Yu X, Gelbič I, Guan X. RAP-PCR fingerprinting reveals time-dependent expression of development-related genes following differentiation process of *Bacillus thuringiensis*. Canadian Journal of Microbiology. 19 Jun 2015;**61**(9):683-690

[38] Ibrahim WA, Ghany A, Nasef SA, Hatem ME. Comparative study on the use of RT-PCR and standard isolation techniques for the detection of *Salmonella* in broiler chicks. International Journal of Veterinary Science and Medicine. 2014;**2**:67

[39] Kawasaki ES. Amplification of RNA. In: Innis MA, Gelfand DH, Stunski JJ, editors. PCR Protocols, a Guide to Methods and Applications. New York: Academic Press; 1990. pp. 21-27

[40] Docker MF, Delvin RH, Richard J, Khattra J. Sensitive and specific PCR-assay detection of Loma salmonae (microsporea). Diseases of Aquatic Organisms. 1997;**29**:41-48

[41] Logan J, Edwards K, Saunders N. Real-Time PCR: Current Technology and Applications. Caister: Academic Press. 2009. ISBN: 978-1-904455-39-4

[42] Espy MJ, Uhl JR, Sloan LM, Buckwalter SP, Jones MF, Vetter EA, et al. Real-time PCR in clinical microbiology: Applications for routine laboratory testing. Clinical Microbiology Reviews. 2006;**19**(1):165-256

[43] Filion M. Quantitative Real-time PCR in Applied Microbiology. Caister: Academic Press; 2012. ISBN: 978-1-908230-01-0

[44] Baldwin BG. Phylogenetic utility of the internal transcribed spacers of nuclear ribosomal DNA in plants: An example from the compositae. Molecular Phylogenetics and Evolution. 1992;**1**(1):3-16

[45] Chou Q, Russell M, Birch DE, Raymond J, Bloch W. Prevention of pre-PCR mis-priming and primer dimerization improves low-copy-number amplifications. Nucleic Acids Research. 1992;**20**(7):1717-1723

[46] Don RH, Cox PT, Wainwright BJ, Baker K, Mattick JS. "Touchdown" PCR to circumvent spurious priming during gene amplification. Nucleic Acids Research. 1991;**19**(14):4008

[47] Stemmer WP, Crameri A, Ha KD, Brennan TM, Heyneker HL. Single-step assembly of a gene and entire plasmid from large numbers of oligodeoxyribonucleotides. Gene. 1995;**164**(1):49-53

[48] Pavlov AR, Pavlova NV, Kozyavkin SA, Slesarev AI. In: Kieleczawa J, editor. DNA Sequencing II: Optimizing Preparation and Cleanup. Jones and Bartlett; 2006. pp. 241-257

[49] Herman JG, Graff JR, Myöhänen S, Nelkin BD, Baylin SB.

Methylation-specific PCR: A novel PCR assay for methylation status of CpG islands. Proceedings of the National Academy of Sciences of the United States of America. 1996;**93**(18):9821-9826

[50] Licchesi JD, Herman JG. Methylation-specific PCR. In: DNA Methylation. Humana Press; 2009. pp. 305-323

[51] Ochman H, Gerber AS, Hartl DL. Genetic applications of an inverse polymerase chain reaction. Genetics. 1988;**120**(3):621-623

[52] Morris T, Robertson B, Gallagher M. Rapid reverse transcription-PCR detection of hepatitis C virus RNA in serum by using the Taq Man fluorogenic detection system. Journal of Clinical Microbiology. 1996;**34**(12):2933-2936

[53] Sheridan GE, Masters CI, Shallcross JA, Mackey BM. Detection of mRNA by reverse transcription-PCR as an indicator of viability in *Escherichia coli* cells. Applied and Environmental Microbiology. 1998;**64**(4):1313-1318

[54] Ruijter JM, Ramakers C, Hoogaars WM, Karlen Y, Bakker O, Van den Hoff MJ, et al. Amplification efficiency: Linking baseline and bias in the analysis of quantitative PCR data. Nucleic Acids Research. 22 Feb 2009;**37**(6):e45

[55] Welsh J, Chada K, Dalal SS, Cheng R, Ralph D, McClelland M. Arbitrarily primed PCR fingerprinting of RNA. Nucleic Acids Research. 11 Sep 1992;**20**(19):4965-4970

[56] Farrar JS, Wittwer CT. Extreme PCR: Efficient and specific DNA amplification in 15-60 seconds. Clinical Chemistry. 1 Jan 2014;**61**(1):145-153

[57] Yu M, Cao Y, Ji Y. The principle and application of new PCR Technologies. In: IOP Conference Series: Earth and Environmental Science; Decemver 2017 (Vol. 100, No. 1, p. 012065). IOP Publishing

[58] Son JH, Cho B, Hong S, Lee SH, Hoxha O, Haack AJ, et al. Ultrafast photonic PCR. Light: Science & Applications. Jul 2015;**4**(7):e280

[59] Wong DK, Tsoi O, Huang FY, Seto WK, Fung J, Lai CL, et al. Application of co-amplification at lower denaturation temperature-PCR (COLD-PCR) sequencing for the early detection of antiviral drug resistance mutations of hepatitis B virus. Journal of Clinical Microbiology. 20 Jun 2014;**52**(9):3209-3215

[60] Liu C, Lin J, Chen H, Shang H, Jiang L, Chen J, et al. Detection of HBV genotypic resistance mutations by coamplification at lower denaturation temperature PCR coupled with Sanger sequencing. Journal of Clinical Microbiology. 4 Jun 2014;**52**(8):2933-2939

[61] Lewandowska MA, Jóźwicki W, Jochymski C, Kowalewski J. Application of PCR methods to evaluate EGFR, KRAS and BRAF mutations in a small number of tumor cells in cytological material from lung cancer patients. Oncology Reports. 1 Sep 2013;**30**(3):1045-1052

[62] Galbiati S, Brisci A, Lalatta F, Seia M, Makrigiorgos GM, Ferrari M, et al. Full COLD-PCR protocol for noninvasive prenatal diagnosis of genetic diseases. Clinical Chemistry. 1 Jan 2011;**57**(1):136-138

[63] Li H, Huang J, Lv J, An H, Zhang X, Zhang Z, et al. Nanoparticle PCR: Nanogold-assisted PCR with enhanced specificity. Angewandte Chemie. 12 Aug 2005;**117**(32):5230-5233

[64] Orpana AK, Ho TH, Alagrund K, Ridanpää M, Aittomäki K, Stenman J. Novel heat pulse extension-PCR-based

method for detection of large CTG-repeat expansions in myotonic dystrophy type 1. The Journal of Molecular Diagnostics. 1 Jan 2013;**15**(1):110-115

[65] Pierce KE, Wangh LJ. Rapid detection and identification of hepatitis C virus (HCV) sequences using mismatch-tolerant hybridization probes: A general method for analysis of sequence variation. BioTechniques. Sep 2013;**55**(3):125-132

[66] Hindson BJ, Ness KD, Masquelier DA, Belgrader P, Heredia NJ, Makarewicz AJ, et al. High-throughput droplet digital PCR system for absolute quantitation of DNA copy number. Analytical Chemistry. 28 Oct 2011;**83**(22):8604-8610

[67] Green MR, Sambrook J, Sambrook J. Molecular Cloning: A Laboratory Manual. 4th ed. New York: Cold Spring Harbor Laboratory Press; 2012

[68] Lakshmi V, Sudha T, Rakhi D, Anilkumar G, Dandona L. Application of polymerase chain reaction to detect HIV-1 DNA in pools of dried blood spots. Indian Journal of Microbiology. 2011;**51**(2):147-152. DOI: 10.1007/s12088-011-0135-0

[69] Valones MAA, Guimarães RL, Brandão LAC, de Souza PRE, de Albuquerque Tavares Carvalho A, Crovela S. Principles and applications of polymerase chain reaction in medical diagnostic fields: A review. Brazilian Journal of Microbiology. 2009;**40**(1):1-11. DOI: 10.1590/S1517-83822009000100001

[70] Novais CM, Pires-Alves M, Silva FF. PCR in real time. Revista Biotecnologia Ciência & Desenvolvimento - Edição nº. 2004;**33**:11

[71] Speers DJ, Ryan S, Harnett G, Chidlow G. Diagnosis of malaria aided by polymerase chain reaction in two cases with low-level parasitaemia. Internal Medicine Journal. Dec 2003;**33**(12):613-615

[72] Estrela C. Metodologia científica: ciência, ensino, pesquisa. Artes Médicas; 2005

[73] Cortelli S, Jorge AO, Querido SM, Cortelli JR. PCR and culture in the subgingival detection of *Actinobacillus actinomycetemcomitans*: A comparative study. Brazilian Dental Science. 12 Aug 2010;**6**(2)

[74] Dwivedi S, Purohit P, Misra R, Pareek P, Goel A, Khattri S, et al. Diseases and molecular diagnostics: A step closer to precision medicine. Indian Journal of Clinical Biochemistry. 1 Oct 2017;**32**(4):374-398

[75] Mahdieh N, Rabbani B. An overview of mutation detection methods in genetic disorders. Iranian Journal of Pediatrics. Aug 2013;**23**(4):375

[76] http://www.genelink.com/gds/GDSintro.asp

[77] Kondo M, Hoshi SL, Yamanaka T, Ishiguro H, Toi M. Economic evaluation of the 21-gene signature (Oncotype DX®) in lymph node-negative/positive, hormone receptor-positive early-stage breast cancer based on Japanese validation study (JBCRG-TR03). Breast Cancer Research and Treatment. 1 Jun 2011;**127**(3):739-749

[78] Klindworth A, Pruesse E, Schweer T, Peplies J, Quast C, Horn M, et al. Evaluation of general 16S ribosomal RNA gene PCR primers for classical and next-generation sequencing-based diversity studies. Nucleic Acids Research. 1 Jan 2013;**41**(1):e1

[79] Marthaler D, Raymond L, Jiang Y, Collins J, Rossow K, Rovira A. Rapid detection, complete genome sequencing, and phylogenetic

analysis of porcine deltacoronavirus. Emerging Infectious Diseases. Aug 2014;**20**(8):1347

[80] Malainey ME. Ancient DNA and the polymerase chain reaction. In: A Consumer's Guide to Archaeological Science. New York, NY: Springer; 2011. pp. 237-253

[81] Wales N, Andersen K, Cappellini E, Ávila-Arcos MC, Gilbert MT. Optimization of DNA recovery and amplification from non-carbonized archaeobotanical remains. PLoS ONE. 27 Jan 2014;**9**(1):e86827

[82] Del Gaudio S, Cirillo A, Di Bernardo G, Galderisi U, Thanassoulas T, Pitsios T, et al. Preamplification procedure for the analysis of ancient DNA samples. The Scientific World Journal. 2013;**2013**

[83] Elion EA, Marina P, Yu L. Constructing recombinant DNA molecules by PCR. Current Protocols in Molecular Biology. 2007 Apr;**78**(1):3-17

[84] Dubey AA, Singh MI, Jain V. Rapid and robust PCR-based all-recombinant cloning methodology. PLoS ONE. 23 Mar 2016;**11**(3):e0152106

[85] Rosen MJ, Callahan BJ, Fisher DS, Holmes SP. Denoising PCR-amplified metagenome data. BMC Bioinformatics. Dec 2012;**13**(1):283

[86] Lee S, Cantarel B, Henrissat B, Gevers D, Birren BW, Huttenhower C, et al. Gene-targeted metagenomic analysis of glucan-branching enzyme gene profiles among human and animal fecal microbiota. The ISME Journal. Mar 2014;**8**(3):493

[87] Rouached H. Efficient procedure for site-directed mutagenesis mediated by PCR insertion of a novel restriction site. Plant Signaling & Behavior. 1 Dec 2010;**5**(12):1547-1548

[88] Edelheit O, Hanukoglu A, Hanukoglu I. Simple and efficient site-directed mutagenesis using two single-primer reactions in parallel to generate mutants for protein structure-function studies. BMC Biotechnology. Dec 2009;**9**(1):61

[89] Cho SH, Jeon J, Kim SI. Personalized medicine in breast cancer: A systematic review. Journal of Breast Cancer. 1 Sep 2012;**15**(3):265-272

[90] Epplen J, Lubjuhnn T. DNA Profiling and DNA Fingerprinting. Springer Science & Business Media; 11 Dec 2012

[91] Mooney C, Raoof R, El-Naggar H, Sanz-Rodriguez A, Jimenez-Mateos EM, Henshall DC. High throughput qPCR expression profiling of circulating microRNAs reveals minimal sex-and sample timing-related variation in plasma of healthy volunteers. PLoS ONE. 23 Dec 2015;**10**(12):e0145316

[92] Sheibani-Tezerji R, Rattei T, Sessitsch A, Trognitz F, Mitter B. Transcriptome profiling of the endophyte Burkholderia phytofirmans PsJN indicates sensing of the plant environment and drought stress. MBio. 30 Oct 2015;**6**(5):e00621-e00615

[93] Condomines M, Hose D, Rème T, Requirand G, Hundemer M, Schoenhals M, et al. Gene expression profiling and real-time PCR analyses identify novel potential cancer-testis antigens in multiple myeloma. The Journal of Immunology. 1 Jan 2009;**183**(2):832-840

[94] Kress WJ, Erickson DL. DNA barcodes: genes, genomics, and bioinformatics. Proceedings of the National Academy of Sciences. 26 Feb 2008;**105**(8):2761-2762

[95] Fraiture MA, Herman P, Taverniers I, De Loose M, Deforce D, Roosens NH. Current and new

approaches in GMO detection: Challenges and solutions. BioMed Research International. 2015;**2015**

[96] Rabiei M, Mehdizadeh M, Rastegar H, Vahidi H, Alebouyeh M. Detection of genetically modified maize in processed foods sold commercially in Iran by qualitative PCR. Iranian Journal of Pharmaceutical Research. 2013;**12**(1):25

[97] Han SH, Oh HS, Cho IC. Identifying the species of origin in commercial sausages in South Korea. Journal of Applied Animal Research. 1 Jan 2017;**45**(1):179-184

[98] Ren J, Deng T, Huang W, Chen Y, Ge Y. A digital PCR method for identifying and quantifying adulteration of meat species in raw and processed food. PLoS ONE. 20 Mar 2017;**12**(3):e0173567

[99] Fumière O, Marien A, Fernández Pierna JA, Baeten V, Berben G. Development of a real-time PCR protocol for the species origin confirmation of isolated animal particles detected by NIRM. Food Additives and Contaminants. 1 Aug 2010;**27**(8):1118-1127

Chapter 3

Genome Engineering for Xenotransplantation

Sean Stevens

Abstract

Xenotransplantation, the transfer of cells, tissues or organs between species, has the potential to overcome the critical need for organs to treat patients. One major barrier in the widespread application of xenotransplantation in the clinic is the overwhelming rejection response that occurs when non-human organs encounter the human immune system. Recent progress in developing new and better genome engineering tools now allows the genetic engineering of genes and pathways in non-human animals to overcome the human rejection response and provide an unlimited supply of rejection-free organs. In this review, the benefits and drawbacks of various genome engineering protocols, and examples of their application in xenotransplantation, are discussed.

Keywords: xenotransplantation, xenoantigen, transgene, gene targeting, gene editing, homologous recombination

1. Introduction

According to the Organ Procurement and Transplantation Network, there are currently over 110,000 patients on the waiting list for organ transplants. Over the past 50 years, therapeutic advances and improvements in surgical techniques have increased the number of patients who could survive and benefit from organ transplantation. Unfortunately, the number of organs available through donation has not changed significantly. Thus, there is a growing disparity between organ supply and demand. Although efforts to enlarge the human donor pool have improved organ availability, even a massive expansion in organ donation would not ensure that a compatible organ would be available when and where necessary for a patient in need. Therefore, other alternatives besides expanded human donation are required.

A variety of efforts are making substantial progress in addressing the lack of organs. One area of research that is rapidly approaching clinical reality is xenotransplantation, the use of animal tissues and organs to treat patients [1]. Aside from the potential for creating an unlimited supply of organs, recent advancements in genome engineering technologies allows the genetic modification of animals to produce donor organs which are less prone to rejection for xenotransplantation in human patients.

2. History

Xenotransplantation experiments were described as early as the seventeenth-century [2], with sporadic attempts made to transplant a variety of animal tissues and

IntechOpen

organs into patients throughout the nineteenth and early twentieth centuries [3–7]. After World War II, organ transplantation from living humans was considered too high risk, and cadaveric organs were insufficient in both quality and number to meet clinical needs. Development of immunosuppressive drugs suggested the possibility that organs from more closely-related mammals could potentially be used in humans. As a consequence, xenotransplantation efforts shifted to the use of organs from primates in human patients. In the 1960s, experiments by Reemstma et al. [8, 9], Hardy et al. [10], and Starzl et al. [11, 12] showed that while it was technically possible to transplant animal organs into humans, there were still too many clinical challenges at that time for the approach to be viable. More research was required to understand and overcome the barriers to the practical application of xenotransplantation in humans.

One of the major advances in xenotransplantation research in the past few decades has been the focus on the use of pigs as donors [13]. This was based, in part, on purely practical considerations. Unlike primates, pigs are an agricultural species for which large scale breeding is well-established. In addition, the evolutionary distance between humans and pigs reduces the risk of transmission of zoonoses from pig organs to patients compared with primate organs. Most importantly, the use of porcine organs does not present the same ethical barriers as the use of non-human primate organs [14].

Although the anatomy and physiology of pig organs is closely analogous to that of humans, the advantages of porcine organ production and availability do not address the critical issue of incompatibilities of non-human tissues and organs with the human immune system [15]. Significant advancements have been made in recent years in understanding the molecular mechanisms of xenorejection responses, and a variety of genetic modifications have been made to overcome these mechanisms. Experiments transplanting pig organs into non-human primates have demonstrated a progressive improvement in organ survival and function as new genetics and drug regimens have been implemented [16]. The FDA is currently developing guidelines for clinical xenotransplantation [17], and efforts to initiate clinical trials in the near term have been announced [18].

3. Immunity and xenotransplantation

The immune system is designed to recognize and eliminate harmful pathogens, while remaining unresponsive to host cells and beneficial microbes. The immune system can be divided into innate and adaptive responses, an interdependent set of activities which both contribute to immunity. The innate response is more immediate, broadly recognizing conserved microbial elements, such as cell wall polysaccharides, and activating a variety of cell types which attack the invading pathogens [19]. The adaptive immune response, which is typically initiated by innate response mechanisms, leads to more precise antigen-specific antibodies and immune cells that continue to control and eliminate pathogens. In addition, the adaptive response creates long-lasting immune "memory" for rapid and specific protection against future infections, as demonstrated by vaccines [20].

Despite being described as separate systems, the innate and adaptive immune responses are highly interdependent and create a layered set of defenses with increasing specificity for pathogens over time [21]. Under normal circumstances, any individual function may not eliminate a given target with 100% efficiency, but when used together in a redundant fashion can prevent nearly all infection. Although the specificity of the immune response indirectly helps to avoid recognition of host tissues, additional tolerance mechanisms are required to restrain the immune system to prevent autoreactivity. Disruptions of the balance between immunity and tolerance can lead to the immune system destroying host tissues (autoimmunity) or allowing repeated severe infections (immunodeficiency) [22].

The transplantation of foreign cells or tissues into a human host can trigger a hostile response from the immune system, leading to immune rejection. The extraordinary precision of the immune system can distinguish even minor differences between donor and recipient, so that even organs from closely related donors may be rejected. Although immunosuppressive drugs can reduce the chance of rejection of human donor organs, the massive amount of immunogenic material found in a whole organ presents an ongoing risk which requires monitoring. Because of the greater genetic differences between pigs and humans, the vigor of the rejection response is much stronger than occurs between human donors and recipients, requiring more and different solutions.

4. Genome engineering and xenotransplantation

A major advantage in using pigs for xenotransplantation is the potential to manipulate the porcine genome to create donor organs that are more compatible for human patients. However, the scope of the engineering challenge in xenotransplantation is extremely large, involving a variety of genes and pathways. With so many potential targets for genetic modification, an assortment of different genome engineering strategies have been applied, including editing or deletion of porcine genes and insertion of human or engineered genes. Because of the great diversity of genome modification efforts being carried out in xenotransplantation research, representative approaches will be highlighted here as examples of the general types of the engineering strategies being employed.

Historically, mice have been subject to more and different genetic modifications than any other mammalian species, and many of the protocols described here were first developed in mice. Aside from their well-established and convenient husbandry, small size, and rapid generation times, mice also have a variety of technological advantages for genome manipulation and production. Although genetic modification has been demonstrated for multiple agricultural species, including pigs, the scale and complexity possible with mice has, until recently, not been available for pigs [23].

One advantage for the creation of mice with multiple genetic modifications is the availability of embryonic stem (ES) cells, which can be cultured *in vitro* for many generations and subject to repeated transfections and selections without loss of competence for production of viable mice [24]. By contrast, pig ES cells have been much more difficult to create, and have not been routinely used for genetic manipulation and production of animals [25]. Cloning of genetically-modified pigs has required use of primary cells, typically fetal, which can be passaged only a short time *in vitro* before losing their competence to produce viable embryos [26]. Therefore, the complex multi-site modifications and selections used in mice are not accessible for use in pigs.

Mouse ES cells not only allow more straightforward and efficient genome engineering, but also facilitate large scale production of cloned mice. The mouse ES cells typically employed for genetic modification can be injected into very early stage embryos (blastocysts) and will aggressively populate the inner cell mass, creating viable chimeric mice which are almost entirely ES-cell derived. Since the ES cells will also contribute to the germ cells of the chimeric mice, the progeny will be highly likely to receive the genetic modifications made to the ES cells [27]. Without readily available porcine ES cells, pig cloning instead relies upon somatic cell nuclear transfer (SCNT), similar to the protocols used to create the sheep

"Dolly". In this approach, pig oocyte nuclei are replaced with nuclei from the modified primary pig cells, and embryonic development stimulated electrochemically. The embryos are transferred to female surrogates and allowed to develop. The level of complexity and effort involved leads to lower efficiencies and higher costs for porcine SCNT relative to mouse ES cell cloning. Additionally, the size and scale of the facilities required for pig cloning is significantly greater compared with mouse cloning, further limiting availability [28].

In the following sections, different types of gene modifications are described with examples of their application in porcine genome engineering for xenotransplantation.

5. Gene deletions

As mentioned above, porcine cells produce molecules which are rapidly recognized by the human immune system and rejected. One straightforward approach to engineering the pig would be to simply eliminate the genes encoding reactive genes by either disrupting or removing the coding sequence. Several of the technical routes which can be employed to accomplish this are discussed below.

5.1 Gene knockout

Gene knockout (KO) approaches developed for use in mouse ES cells generally rely upon homologous recombination to replace a region of genomic DNA with a heterologous DNA sequence, which interrupts the function of the target gene [29]. To accomplish this, a DNA vector is generated with the heterologous DNA flanked on either side by sequences identical to regions flanking the genomic region to be eliminated. When introduced into mouse ES cells, the flanking sequences of the DNA vector first align with the cognate regions of the genome on either side of the target gene, after which the homologous recombination machinery replaces the genomic target with the heterologous DNA vector sequence found between the flanking sequences.

Because homologous recombination occurs at a relatively low rate, in order to identify properly targeted cells within the larger cell population, it is common to include a gene in the heterologous DNA to be inserted into the genome, which, once properly inserted, allows selection of the desired cells. For example, genes which confer resistance to drugs which kill mammalian cells (neomycin hygromycin or puromycin resistance), or genes encoding molecules that enable cells to be isolated via flow cytometry (green fluorescent protein or novel cell surface markers), allow isolation of even extremely rare targeted cells from a large mixed population [30].

The use of gene KO approaches was one of the earliest successes in pig genetic modification for xenotransplantation [31]. The porcine genome encodes proteins that can be substantially different from their human counterparts, or that carry additional modifications which are not present in humans and can induce immune responses. These molecules are collectively referred to as "xenoantigens" [32]. Some of the most reactive of these targets are carbohydrate molecules found as post-translational modifications to proteins observed in pig, but not human, cells. Human serum can contain high levels of pre-existing antibodies specific for these porcine-specific glycan epitopes, leading to the destruction of pig cells expressing these molecules through antibody-dependent cell-mediated cytotoxicity (ADCC) or complement-dependent cytotoxicity (CDC) mechanisms [33]. It is not entirely clear why human serum carries antibodies to these particular carbohydrates; one proposal is that the xenoantigens are related to glycans found in the cell walls of pathogens, others suggest that the human dietary consumption of pork causes antibody generation to the porcine-specific molecules [34]. Because the novel carbohydrate structures are created by specific glycosylation enzymes, it is possible to eliminate the gene responsible for the enzymatic activity and prevent the xenoantigen from being expressed by pig cells.

The GGTA1 gene encodes the enzyme responsible for creating the highly reactive glycan Gal alpha (1,3) Gal epitope in pigs [35]. The KO of the GGTA1 gene is one of the earliest genetic modifications of pigs for application in xenotransplantation, and resulted in greatly reduced human antibody recognition of porcine cells [36, 37]. However, ablation of the GGTA1 gene alone did not completely eliminate porcine cell recognition by human serum antibodies. The enzymes responsible for other xenoantigens, such as CMAH (cytidine monophosphate-N-acetylneuraminic acid hydroxylase critical for Neu5Gc biosynthesis) and B4GALNT2 (beta 1,4 N-acetylgalactosaminyltran sferase), have been identified as sources of porcine-specific epitopes bound by antibodies found in human serum. In each case, the deletion of the gene responsible for creating the specific glycan leads to greatly decreased recognition of porcine cells by antibodies in human serum, and reduction in complement-mediated destruction [38, 39].

Another subset of xenoantigens is the swine leukocyte antigens (SLA), the physical and functional equivalent of the human leukocyte antigens (HLA) [40]. Much like the case for human HLA, the SLA genes are highly diverse and individual patients will have a variable level of cross-reactive antibodies in their serum for a given set of SLA genes [41]. Although typing of patients and porcine donors to find the best HLA-SLA matches would be similar to the current system used for determining allotransplant cross-reactivity [42], use of gene targeting or editing technologies could easily eliminate the genes encoding SLA entirely. However, unlike the glycan epitopes described above, the SLA have a critical role in antigen presentation as part of the immune response, and thus the deletion of SLA could create risks of immune deficiencies that outweigh their risks as xenoantigens. Instead, alternate approaches seek to create engineered SLA proteins lacking the epitopes responsible for the immunogenicity while maintaining their antigen presentation functions [43].

5.2 Gene editing

The ease and efficiency of creating gene KO has improved recently through the use of engineered molecules to create genome disruptions in a process referred to as "gene editing". These novel molecules can be designed to generate double-strand DNA breaks at virtually any chosen genomic site *in situ*. Cellular machinery closely surveils the genome for double-strand breaks which are then recognized and often repaired by non-homologous end joining (NHEJ). Because NHEJ relies upon small single-strand overlaps at the ends of a break, the repair may be imprecise and, if within a coding region, can lead to frame shift mutations which inactivate the gene [44].

The most prominent of these novel tools for gene editing are Zinc Finger Nucleases (ZFN), Transcription Activator-Like Effector Nuclease (TALEN) and Clustered Regularly Interspaced Short Palindromic Repeats (CRISPR), each of which consists of two regions: a sequence-specific DNA binding domain and an enzymatic function that creates a double-strand break in the target DNA [45–47]. For ZFN and TALEN, the synthetic DNA binding domain is created by repetitive protein modules which can be joined combinatorially to recognize a particular DNA sequence. Both approaches, while successful, require a significant investment of time and resources to identify functional molecules. CRISPR, like ZFN and TALEN, has the ability to generate double-strand DNA breaks, however, the DNA binding domain relies upon RNA base-pairing with target DNA for its precision. The use of an RNA to guide specificity greatly improves the speed and efficiency in identifying optimal molecules at a much lower cost, which has led to its rapid adoption in genome engineering [48, 49].

The use of CRISPR for the rapid modification of the pig genome was recently demonstrated with the ablation of porcine endogenous retroviral (PERV) sequences. The pig genome carries 25 or more copies of these gamma retroviral sequences, which are transmitted from parent to offspring through inheritance. Application of CRISPR was able to eliminate the PERV sequences from the genome of porcine cells [50]. Although the potential risk of infectious disease from porcine organs caused by PERV sequences in xenotransplantation is debatable, the results show the ability of CRISPR to target multiple, homologous loci throughout the genome. A key question that has arisen regarding large scale CRISPR targeting at multiple genomic sites is whether significant numbers of off-target double-stranded breaks were introduced, which may create unexpected mutations in the resulting pigs [51]. Nonetheless, the ease of use of CRISPR has resulted in widespread adoption for genome engineering in xenotransplantation.

Gene deletion has been instrumental in the advancement of xenotransplantation, however, there are limitations to its application; the genes of interest must be non-essential to pig viability, development, fertility and, most importantly, organ function. The number of distinct loci to be targeted is also a serious consideration, since independently-assorting alleles will be challenging to breed together in a reasonable timeframe. To address these concerns, additional engineering strategies are required as discussed below.

6. Gene insertions

Gene KO and editing techniques have been used for ablation of xenoantigen genes but do not address the need to express human or synthetic genes in pig cells and organs. Unlike gene deletions, gene insertions require heterologous DNA to be introduced into the genome in a manner that allows subsequent expression of the gene(s) encoded by the inserted DNA. Because they are being transferred into the genome from another source, these novel genes are referred to as "transgenes" (TG), whether they are derived from natural or synthetic sequences. The general approaches to introduce TG into the genome are detailed below.

6.1 Random integrant TG

One of the earliest types of genetic modification described in mammals was insertion of DNA into the target genome by random integration. After transfection of DNA into nearly all mammalian cells, some portion of the heterologous DNA can be found incorporated at random sites in the genome [52]. The precise process for this is unclear, but presumably is a result of aberrant repair mechanisms. One hypothesis is that endogenous NHEJ machinery recognizes breaks in the genome and fortuitously utilizes the relatively higher concentration of the heterologous DNA vector sequences to repair the break [53], resulting in the insertion of the TG into the genome.

Agricultural species, including pig, were some of the earliest TG animals described, establishing the utility of this approach [54]. For xenotransplantation, several of the initial TG approaches focused on inhibiting human antibody-mediated damage of porcine organs. The binding of human antibodies to porcine cells leads to complement pathway activation and subsequent cell ablation [55]. The complement function is controlled by several proteins, such as CD46, CD55 and CD59, referred to as complement regulatory proteins (CRPs). The CRPs are broadly expressed on many different cell types to prevent harm from complement activity by raising the threshold of antibody binding required for complement pathway induction [56]. The transgenic expression of human CRPs in pigs appears to overcome human complement activity, and may have the potential to reduce, or even eliminate, the need for xenoantigen KOs. By placing the human CRP TGs under the control of strong gene expression elements, the CRP protein levels on the porcine cells can be much higher than CRP levels on normal human cells, further increasing resistance to complement-mediated destruction [57].

Pig lineages developed by multiple labs have been engineered to express human CRPs, individually or in combination. In most cases, the porcine cells appeared to be more resistant to complement-mediated destruction, and organs from TG animals survived longer in xenotransplant experiments in non-human primates [58–60]. Because each of the CRPs control a different part of the complement pathway, the use of multiple human TGs was more effective in protecting cells from complement-mediated destruction than individual TGs [61]. Together with the removal of key xenoantigens, the expression of human CRPs by porcine cells has greatly reduced the effects of human serum antibodies on xenografts.

Because random insertion of DNA does not require homologous recombination, it is relatively rapid and efficient to produce transgenic animals [62]. The process is so efficient that the selection methods that are critical for gene KO described above are often unnecessary for TG. Despite the speed and ease of generation, random integration of TGs has several drawbacks. Variabilities in genome structure can affect the expression level of genes inserted at distinct chromosomal regions, such that identical TGs may express at very different levels depending upon their specific location [63]. Furthermore, multiple copies of a given TG may be inserted into the genome at a single site, creating concatenated repeats which can be unstable and yield variable expression levels [64]. Random TG insertion may occur within or near endogenous genes and alter or inactivate their function, leading to tumorigenesis, instability or even lethality [65].

6.2 Homologous recombination

The development of techniques for precise gene KO by homologous recombination has been adapted for site-specific gene insertion or gene knock-in (KI). Similar to the KO vectors described above, the gene to be introduced is flanked by DNA sequences that are identical to regions of the genome to be targeted. After introduction of the heterologous DNA vector, the regions of DNA sequence identity are aligned with the target genome sequence, after which the homologous recombination machinery catalyzes reactions which swap the endogenous genomic DNA with the heterologous DNA within the construct. If the recombination event occurs with high fidelity, the gene of interest will functionally replace the gene that was removed [24]. Similar to gene KO, this approach is much less efficient than random TG integration. Therefore, vectors carrying the TG are often designed to incorporate selectable markers, similar to those used for gene KO, to allow the identification of cells carrying the desired TG in the genome. In this case, both targeted and randomly integrated TGs may be selected, requiring additional assays, such as PCR or Southern blotting, to distinguish between sequence-specific and random insertion events [30].

As described above for gene KO, specific targeting is more efficient in murine ES cells, which express the enzymatic machinery necessary for homologous recombination, than is currently possible for pig primary cells. Insertion of heterologous DNA into the mammalian genome is believed to be driven by endogenous DNA repair mechanisms, presumably in response to DNA breaks, whether randomly via NHEJ, or specifically via homologous recombination [66]. The deliberate introduction of double-stranded DNA breaks at the desired integration site should therefore improve the efficiency of heterologous DNA insertion by activating and recruiting the cellular repair machinery. Application of ZFN, TALEN and CRISPR technologies have shown that homologous recombination efficiencies are improved when one or more double-strand DNA breaks are introduced into the genome at the desired site of insertion [67] with CRISPR exhibiting bi-allelic targeting rates as high as 90% [68]. The use of these more advanced genome engineering tools has greatly improved the rates and specificity of both gene deletion and gene insertion in genomes.

Gene insertion by homologous recombination for xenotransplantation has not advanced as rapidly as other approaches, in part due to the challenges of using primary porcine cells, for which the efficiencies can be extremely low, particularly with large DNA constructs. Use of improved genome engineering tools with increased targeting efficiencies have already been applied in pig and will continue to grow in impact [69]. However, even with improved efficiencies of gene insertion, breeding pigs with multiple, independently segregating loci is challenging. The number of litters required to produce animals bearing all of the genetic modifications greatly increases with each additional locus, which can be impractical for large animals such as pigs. Therefore, the ability to insert the maximal amount of genetic information into the genome in the minimal number of steps, as discussed below, is highly valuable.

6.3 Multigenic insertion

The most straightforward example of multigenic targeting at a single site takes advantage of the observation that multiple DNA vectors co-transfected into cells will tend to insert together at a given genome site. This approach was used successfully to simultaneously introduce as many as five large transgenes in a single step into porcine cells [70]. Although difficulties in producing mature cloned pigs limited the study to animals with fewer integrated genes, the study demonstrated the feasibility of rapidly making animals with multiple TG.

Another, relatively less complicated, way to introduce multiple TG is to generate large DNA constructs bearing multiple TGs for integration at random into the genome [71]. This greatly reduces the complexity of screening, while increasing the efficiency of insertion, but still relies upon random integrants which can have variable TG copy numbers and expression levels.

Multiple genomic regions have been defined, such as Rosa26, which allow expression of heterologous genes at similar levels regardless of cell type. These "safe harbor" regions are believed to have a chromatin structure that is more easily accessible to the gene expression machinery, regardless of cell type. Targeting at porcine homologs of murine safe harbor sites such as Rosa26 has been described and demonstrates the utility of this approach [70, 72]. On a practical level, the use of safe harbor sites yields more reproducible gene expression than random TG insertions, so fewer lineages are required to select animals with desired TG levels. Furthermore, the defined location and copy number of TG inserted at a safe harbor site makes breeding and genotyping more straightforward, and is expected to provide a less complicated regulatory pathway for clinical use.

As DNA synthesis and assembly has improved, increasingly large DNA constructs encoding a variety of TG are possible, however, as the size of the DNA increases, the rate of insertion decreases. Considering the lower insertion rates observed for large animals such as pig, alternate approaches are necessary to incorporate larger DNAs into the genome.

6.4 Site-specific recombination

Bacteriophage- and yeast-derived site-specific recombinases are, as the name suggests, proteins which catalyze recombination between two specific DNA recognition sites, small (<50 base pair) sequences that are unique to the recombinase being used. The recombination event is highly efficient, in some cases eliminating the need for selection genes, and allowing large DNA constructs to be inserted at a much higher frequency than possible for homologous recombination [73].

One limitation for the use of site-specific recombination is the need for a recognition sequence to be present in the target genome at the desired locus. This requires a preceding step in which the recognition site is engineered into the genome using less efficient homologous recombination. Therefore, the gain in efficiency for introduction of large DNA constructs may be offset by the need for insertion of the recombinase recognition site into the genome. Despite this constraint, the potential for site-specific recombination into a defined locus has been demonstrated in pigs [70] and provides a route for more rapid complex genetic modifications.

7. Future needs

The advancements in genome engineering, both in general and in their application to xenotransplantation, have been significant, but many needs remain to be addressed. As new genome engineering tools are identified and further refined, improvement of targeting efficiencies will allow more sophisticated modifications of the pig genome. Ideally, the pig genome will become as readily manipulated as the mouse, allowing researchers to further leverage approaches shown to be effective in murine models.

One major technological difference in the genetic modification of mice and pigs (and many other mammalian species) is the lack of ES cells possessing significant rates of homologous recombination that can be grown in culture for extended periods and subjected to multiple manipulations without losing the ability to produce viable pigs. Efforts to identify natural or induced pluripotent stem cells (iPSC) suitable for these purposes have been described, but have yet to demonstrate practical application for porcine genome engineering [74]. Ongoing work will be required to identify and validate cells which meet these needs.

The function of the TGs themselves can also be further improved. The majority of TG constructs used in pigs have used constitutive promoters to drive high level expression of the proteins encoded by the TGs. In some cases, such as CRPs, this approach may be useful, however, overexpression of TGs which inhibit critical immune processes may create risks of immune deficiency and infections. For this reason, use of expression control elements which can turn on and off TG activity is of increasing interest in xenotransplantation. There are multiple examples of inducible promoters employed in mice which can be controlled by exogenously applied small molecules (such as the tetracycline repressor system), or by endogenous signals (such as promoters for innate immune response genes) [75]. Advanced DNA synthesis and assembly methods also allow synthetic biology approaches to create novel signaling pathways and networks not present in nature.

Immune tolerance is another very active area in xenotransplantation research. As the molecular mechanisms controlling the balance between immunity and tolerance are further elucidated, manipulation of the human immune system itself to specifically reduce or eliminate responses to porcine targets, while leaving intact immunity to infectious diseases, will help overcome xenorejection. Multiple approaches are currently being tested and genome targets identified to encourage human immune tolerance of porcine cells and tissues [76, 77].

8. Conclusions

The speed and ease of genome engineering technologies has helped to overcome many of the limitations for the use of pig organs for xenotransplantation. Despite recent achievements, a key question remains: which combination of genetic modifications is most critical to make a pig organ useful for xenotransplantation? Ongoing experiments seek to address this question, but the answers are likely to be complex and dependent upon the type of organ, the specific immune mechanisms involved, and perhaps other factors that are not yet defined. It is very likely that the first set of genetic modifications of pigs used for xenotransplantation in humans will not be the final set, as the understanding of the mechanisms of xenorejection increases and better strategies developed to influence the human immune response. Continuing progress in genome engineering technologies of pigs will allow the creation of the more complex modifications necessary to meet these demands. Although much remains to be done, it is clear that given the current rate of progress, overcoming the crisis of human organ shortage with unlimited rejection-free porcine organs is rapidly growing closer to reality.

Author details

Sean Stevens
Synthetic Genomics Inc., CA, USA

*Address all correspondence to: sean_m_stevens@yahoo.com

References

[1] Zeyland J, Lipiński D, Słomski R. The current state of xenotransplantation. Journal of Applied Genetics. 2015;**56**(2): 211-218. DOI: 10.1007/s13353-014-0261-6

[2] Roux FA, Sai P, Deschamps JY. Xenotransfusions, past and present. Xenotransplantation. 2007;**14**(3):208-216. DOI: 10.1111/j.1399-3089.2007.00404.x

[3] Gibson T. Zoografting: The early, early years. British Journal of Plastic Surgery. 1955;**8**(3):234-242

[4] Princeteau M. Greffe renale. Journal de Medecine de Bordeaux. 1905;**26**:549

[5] Jaboulay M. Greffe de reins au pli coude par soudres artielles et veincuses. Lyon Médical. 1906;**107**:575

[6] Unger E. Nierentransplantation. Klinische Wochenschrift. 1910;**47**:573

[7] Neuhof H. The Transplantation of Tissues. New York: Appleton and Co; 1923. p. 260

[8] Reemstma K, McCracken BH, Schlegel JU, Pearl MA, Pearce CW, DeWitt CW, et al. Renal transplantation in man. Annals of Surgery. 1964;**160**:384

[9] Hume D. Discussion of paper by Reemtsma and others. Annals of Surgery. 1964;**160**:384

[10] Hardy JD, Chavez CM, Kurrus FD, Neely WA, Erasian S, Turner MD, et al. Heart transplantation in man. Journal of the American Medical Association. 1964;**188**:1132

[11] Starzl TE, Marchioro TL, Peters GN, Kirkpatrick CH, Wilson WE, Porter KA, et al. Renal heterotransplantation from baboon to man: Experience with 6 cases. Transplantation. 1964;**2**:752-776

[12] Starzl TE, Fung J, Tzakis A, Todo S, Demetris AJ, Marino IR, et al. Baboon-to-liver transplantation. Lancet. 1993;**341**(8837):65-71

[13] Klymiuk N, Aigner B, Brem G, Wolf E. Genetic modification of pigs as organ donors for xenotransplantation. Molecular Reproduction and Development. 2010;**77**(3):209-221. DOI: 10.1002/mrd.21127

[14] Shapiro RS. Future issues in transplantation ethics: Ethical and legal controversies in xenotransplantation, stem cell, and cloning research. Transplantation Reviews (Orlando, FL.). 2008;**22**(3):210-214. DOI: 10.1016/j.trre.2008.04.004

[15] Cooper DK, Gollackner B, Sachs DH. Will the pig solve the transplantation backlog? Annual Review of Medicine. 2002;**53**:133-147

[16] Yung GLP, Rieben R, Buhler L, Schuurman HJ, Seebach JD. Xenotransplantation: where do we stand in 2016? Swiss Medical Weekly. 2017;**147**:w14403. DOI: 10.4414/smw.2017.14403

[17] Cooper DKC, Pierson RN, Hering BJ, Mohiuddin MM, Fishman JA, Denner J, et al. Regulation of clinical xenotransplantation-time for a reappraisal. Transplantation. 2017;**101**(8):1766-1769. DOI: 10.1097/TP.0000000000001683

[18] Pullen LC. Xenotransplantation: Time to get excited? American Journal of Transplantation. 2017;**17**(12): 2995-2996. DOI: 10.1111/ajt.14553

[19] Riera Romo M, Pérez-Martínez D, Castillo Ferrer C. Innate immunity in vertebrates: An overview. Immunology. 2016;**148**(2):125-139. DOI: 10.1111/imm.12597

[20] Bonilla FA, Oettgen HC. Adaptive immunity. The Journal of Allergy and

Clinical Immunology. 2010;**125** (2 Suppl 2):S33-S40. DOI: 10.1016/j.jaci.2009.09.017

[21] Schenten D, Medzhitov R. The control of adaptive immune responses by the innate immune system. Advances in Immunology. 2011;**109**:87-124. DOI: 10.1016/B978-0-12-387664-5.00003-0

[22] Delves PJ, Martin SJ, Burton DR, Roitt IM. Roitt's Essential Immunology. United Kingdom: Wiley-Blackwell; 2017

[23] van der Weyden L, White JK, Adams DJ, Logan DW. The mouse genetics toolkit: Revealing function and mechanism. Genome Biology. 2011;**12**(6):224. DOI: 10.1186/gb-2011-12-6-224

[24] Capecchi MR. Gene targeting in mice: Functional analysis of the mammalian genome for the twenty-first century. Nature Reviews. Genetics. 2005;**6**(6):507-512. DOI: 10.1038/nrg1619

[25] Wheeler MB. Development and validation of swine embryonic stem cells: A review. Reproduction, Fertility, and Development. 1994;**6**(5):563-568

[26] Keefer CL. Artificial cloning of domestic animals. Proceedings of the National Academy of Sciences of the United States of America. 2015;**112**(29):8874-8878. DOI: 10.1073/pnas.1501718112

[27] Poueymirou WT, Auerbach W, Frendewey D, Hickey JF, Escaravage JM, Esau L, et al. F0 generation mice fully derived from gene-targeted embryonic stem cells allowing immediate phenotypic analyses. Nature Biotechnology. 2007;**25**(1):91-99. DOI: 10.1038/nbt1263

[28] Wilmut I, Beaujean N, de Sousa PA, Dinnyes A, King TJ, Paterson LA, et al. Somatic cell nuclear transfer. Nature. 2002;**419**(6907):583-586. DOI: 10.1038/nature01079

[29] Bradley A, Ramírez-Solis R, Zheng H, Hasty P, Davis A. Genetic manipulation of the mouse via gene targeting in embryonic stem cells. Ciba Foundation Symposium. 1992;**165**:256-269

[30] Clarke AR. Transgenesis Techniques. Totowa, NJ: Humana Press; 2002

[31] Tai HC, Ezzelarab M, Hara H, Ayares D, Cooper DK. Progress in xenotransplantation following the introduction of gene-knockout technology. Transplant International. 2007;**20**(2):107-117. DOI: 10.1111/j.1432-2277.2006.00398.x

[32] Cooper DK. Xenoantigens and xenoantibodies. Xenotransplantation. 1998;**5**(1):6-17

[33] Galili U. The α-Gal epitope (Galα1-3Galβ1-4GlcNAc-R) in xenotransplantation. Biochimie. 2001;**83**(7):557-563

[34] Samraj AN, Pearce OM, Läubli H, Crittenden AN, Bergfeld AK, Banda K, et al. A red meat-derived glycan promotes inflammation and cancer progression. Proceedings of the National Academy of Sciences of the United States of America. 2015;**112**(2):542-547. DOI: 10.1073/pnas.1417508112

[35] Galili U. The alpha-gal epitope and the anti-Gal antibody in xenotransplantation and in cancer immunotherapy. Immunology and Cell Biology. 2005;**83**(6):674-686. DOI: 10.1111/j.1440-1711.2005.01366.x

[36] Phelps CJ, Koike C, Vaught TD, Boone J, Wells KD, Chen SH, et al. Production of alpha 1,3-galactosyltransferase-deficient pigs. Science. 2003;**299**(5605):411-414. DOI: 10.1126/science.1078942

[37] Lai L, Kolber-Simonds D, Park KW, Cheong HT, Greestein JL, Im GS, et al.

Production of alpha-1,3-galactosyltransferase knockout pigs by nuclear transfer cloning. Science. 2002;**295**(5557):1089-1092. DOI: 10.1126/science.1068228

[38] Estrada JL, Martens G, Li P, Adams A, Newell KA, Ford ML, et al. Evaluation of human and non-human primate antibody binding to pig cells lacking GGTA1/CMAH/β4GalNT2 genes. Xenotransplantation. 2015;**22**(3):194-202. DOI: 10.1111/xen.12161

[39] Cimeno A, Hassanein W, French BM, Powell JM, Burdorf L, Goloubeva O, et al. N-glycolylneuraminic acid knockout reduces erythrocyte sequestration and thromboxane elaboration in an ex vivo pig-to-human xenoperfusion model. Xenotransplantation. 2017;24(6):e12339. DOI: 10.1111/xen.12339

[40] Lunney JK, Ho CS, Wysocki M, Smith DM. Molecular genetics of the swine major histocompatibility complex, the SLA complex. Developmental and Comparative Immunology. 2009;**33**(3):362-374. DOI: 10.1016/j.dci.2008.07.002

[41] Barreau N, Godfrin Y, Bouhours JF, Bignon JD, Karam G, Leteissier E, et al. Interaction of anti-HLA antibodies with pig xenoantigens. Transplantation. 2000;**69**(1):148-156

[42] Martens GR, Reyes LM, Butler JR, Ladowski JM, Estrada JL, Sidner RA, et al. Humoral reactivity of renal transplant-waitlisted patients to cells from GGTA1/CMAH/B4GalNT2, and SLA class I knockout pigs. Transplantation. 2017;**101**(4):e86-e92. DOI: 10.1097/TP.0000000000001646

[43] Ladowski JM, Martens GR, Reyes LM, Wang ZY, Eckhoff DE, Hauptfeld-Dolejsek V, et al. Examining the biosynthesis and xenoantigenicity of class II swine leukocyte antigen proteins. Journal of Immunology. 2018;**200**(8):2957-2964. DOI: 10.4049/jimmunol.1800022

[44] Chang HHY, Pannunzio NR, Adachi N, Lieber MR. Non-homologous DNA end joining and alternative pathways to double-strand break repair. Nature Reviews. Molecular Cell Biology. 2017;**18**(8):495-506. DOI: 10.1038/nrm.2017.48

[45] Urnov FD, Rebar EJ, Holmes MC, Zhang HS, Gregory PD. Genome editing with engineered zinc finger nucleases. Nature Reviews. Genetics. 2010;**11**(9):636-646. DOI: 10.1038/nrg2842

[46] Mahfouz MM, Li L, Shamimuzzaman M, Wibowo A, Fang X, Zhu JK. De novo-engineered transcription activator-like effector (TALE) hybrid nuclease with novel DNA binding specificity creates double-strand breaks. Proceedings of the National Academy of Sciences of the United States of America. 2011;**108**(6):2623-2628. DOI: 10.1073/pnas.1019533108

[47] Eid A, Mahfouz MM. Genome editing: The road of CRISPR/Cas9 from bench to clinic. Experimental & Molecular Medicine. 2016;**48**(10):e265. DOI: 10.1038/emm.2016.111

[48] Kim JS. Genome editing comes of age. Nature Protocols. 2016;**11**(9):1573-1578. DOI: 10.1038/nprot.2016.104

[49] Gaj T, Gersbach CA, Barbas CF. ZFN, TALEN, and CRISPR/Cas-based methods for genome engineering. Trends in Biotechnology. 2013;**31**(7):397-405. DOI: 10.1016/j.tibtech.2013.04.004

[50] Niu D, Wei HJ, Lin L, George H, Wang T, Lee IH, et al. Inactivation of porcine endogenous retrovirus in pigs using CRISPR-Cas9. Science. 2017;**357**(6357):1303-1307. DOI: 10.1126/science.aan4187

[51] Kadam US, Shelake RM, Chavhan RL, Suprasanna P. Concerns regarding 'off-target' activity of genome editing endonucleases. Plant Physiology and Biochemistry. 2018;**131**:22-30. DOI: 10.1016/j.plaphy.2018.03.027

[52] Smith K. Theoretical mechanisms in targeted and random integration of transgene DNA. Reproduction, Nutrition, Development. 2001;**41**(6):465-485

[53] Rodgers K, McVey M. Error-prone repair of DNA double-strand breaks. Journal of Cellular Physiology. 2016;**231**(1):15-24. DOI: 10.1002/jcp.25053

[54] Pursel VG, Pinkert CA, Miller KF, Bolt DJ, Campbell RG, Palmiter RD, et al. Genetic engineering of livestock. Science. 1989;**244**(4910):1281-1288

[55] Dalmasso AP. The complement system in xenotransplantation. Immunopharmacology. 1992;**24**(2):149-160

[56] Kim DD, Song WC. Membrane complement regulatory proteins. Clinical Immunology. 2006;**118**(2-3):127-136. DOI: 10.1016/j.clim.2005.10.014

[57] McCurry KR, Kooyman DL, Alvarado CG, Cotterell AH, Martin MJ, Logan JS, et al. Human complement regulatory proteins protect swine-to-primate cardiac xenografts from humoral injury. Nature Medicine. 1995;**1**(5):423-427

[58] McGregor CG, Davies WR, Oi K, Teotia SS, Schirmer JM, Risdahl JM Tazelaar HD, et al. Cardiac xenotransplantation: Recent preclinical progress with 3-month median survival. The Journal of Thoracic and Cardiovascular Surgery. 2005;**130**(3): 844-851. DOI: 10.1016/j.jtcvs.2005.04.017

[59] Cozzi E, Bhatti F, Schmoeckel M, Chavez G, Smith KG, Zaidi A, et al. Long-term survival of nonhuman primates receiving life-supporting transgenic porcine kidney xenografts. Transplantation. 2000;**70**(1):15-21

[60] Niemann H, Verhoeyen E, Wonigeit K, Lorenz R, Hecker J, Schwinzer R, et al. Cytomegalovirus early promoter induced expression of hCD59 in porcine organs provides protection against hyperacute rejection. Transplantation. 2001;**72**(12):1898-1906

[61] Zhou CY, McInnes E, Copeman L, Langford G, Parsons N, Lancaster R, et al. Transgenic pigs expressing human CD59, in combination with human membrane cofactor protein and human decay-accelerating factor. Xenotransplantation. 2005;**12**(2):142-148. DOI: 10.1111/j.1399-3089.2005.00209.x

[62] Ittner LM, Götz J. Pronuclear injection for the production of transgenic mice. Nature Protocols. 2007;**2**(5):1206-1215. DOI: 10.1038/nprot.2007.145

[63] Laboulaye MA, Duan X, Qiao M, Whitney IE, Sanes JR. Mapping transgene insertion sites reveals complex interactions between mouse transgenes and neighboring endogenous genes. Frontiers in Molecular Neuroscience. 2018;**11**:385. DOI: 10.3389/fnmol.2018.00385

[64] Zhou CY, McInnes E, Copeman L, Langford G, Parsons N, Lancaster R, Richards A, Carrington C, Thompson S. Transgenic pigs expressing human CD59, in combination with human membrane cofactor protein and human decay-accelerating factor. Xenotransplantation. 2005;**12**(2):142-148. DOI: 10.1111/j.1399-3089.2005.00209.x

[65] Carlson CM, Largaespada DA. Insertional mutagenesis in mice: New perspectives and tools. Nature Reviews. Genetics. 2005;**6**(7):568-580. DOI: 10.1038/nrg1638

[66] Hasty P, Rivera-Pérez J, Bradley A. The role and fate of DNA ends for homologous recombination in embryonic stem cells. Molecular and Cellular Biology. 1992;**12**(6):2464-2474

[67] Wells KD, Prather RS. Genome-editing technologies to improve research, reproduction, and production in pigs. Molecular Reproduction and Development. 2017;**84**(9):1012-1017. DOI: 10.1002/mrd.22812

[68] Acosta S, Fiore L, Carota IA, Oliver G. Use of two gRNAs for CRISPR/Cas9 improves bi-allelic homologous recombination efficiency in mouse embryonic stem cells. Genesis. 2018;**56**(5):e23212. DOI: 10.1002/dvg.23212

[69] Ryu J, Prather RS, Lee K. Use of gene-editing technology to introduce targeted modifications in pigs. Journal of Animal Science and Biotechnology. 2018;**9**:15. DOI: 10.1186/s40104-017-0228-7

[70] Fischer K, Kraner-Scheiber S, Petersen B, Rieblinger B, Buermann A, Flisikowska T, et al. Efficient production of multi-modified pigs for xenotransplantation by 'combineering', gene stacking and gene editing. Scientific Reports. 2016;**6**:29081. DOI: 10.1038/srep29081

[71] Rostovskaya M, Naumann R, Fu J, Obst M, Mueller D, Stewart AF, et al. Transposon mediated BAC transgenesis via pronuclear injection of mouse zygotes. Genesis. 2013;**51**(2):135-141. DOI: 10.1002/dvg.22362

[72] Li X, Yang Y, Bu L, Guo X, Tang C, Song J, et al. Rosa26-targeted swine models for stable gene over-expression and Cre-mediated lineage tracing. Cell Research. 2014;**24**(4):501-504. DOI: 10.1038/cr.2014.15

[73] Hallet B, Sherratt DJ. Transposition and site-specific recombination: Adapting DNA cut-and-paste mechanisms to a variety of genetic rearrangements. FEMS Microbiology Reviews. 1997;**21**(2):157-178. DOI: 10.1111/j.1574-6976.1997.tb00349.x

[74] Secher JO, Callesen H, Freude KK, Hyttel P. Initial embryology and pluripotent stem cells in the pig—The quest for establishing the pig as a model for cell therapy. Theriogenology. 2016;**85**(1):162-171. DOI: 10.1016/j.theriogenology.2015.09.017

[75] Jin YX, Jeon Y, Lee SH, Kwon MS, Kim T, Cui XS, et al. Production of pigs expressing a transgene under the control of a tetracycline-inducible system. PLoS One. 2014;**9**(1):e86146. DOI: 10.1371/journal.pone.0086146

[76] Vagefi PA, Shah JA, Sachs DH. Progress towards inducing tolerance of pig-to-primate xenografts. International Journal of Surgery. 2015;**23**(Pt B):291-295. DOI: 10.1016/j.ijsu.2015.07.720

[77] Griesemer A, Yamada K, Sykes M. Xenotransplantation: Immunological hurdles and progress toward tolerance. Immunological Reviews. 2014;**258**(1):241-258. DOI: 10.1111/imr.12152

Chapter 4

Prospects for the Production of Recombinant Therapeutic Proteins and Peptides in Plants: Special Focus on Angiotensin I-Converting Enzyme Inhibitory (ACEI) Peptides

Carolina Gomes, Filipe Oliveira, Sandra Isabel Vieira and Ana Sofia Duque

Abstract

Molecular pharming is a cost-effective, scalable, and safe system to produce high-quality and biologically active recombinant therapeutic proteins. Thus, plants are emerging alternative platform for the production of pharmaceutically relevant proteins such as vaccines, antibodies, antibody derivatives, and some serum-derived proteins. Additionally, plants have also been used to produce bioactive and immunogenic peptides. The efficacy, selectivity, specificity, and low toxicity make them particularly well-suited therapeutic agents for various indications, for instance, cardiovascular and infectious diseases, immunological disorders, and cancer. In the broad range of known bioactive peptides, angiotensin I-converting enzyme inhibitory (ACEI) peptides derived from food proteins have attracted particular attention for their ability to prevent hypertension. So far, several ACEI peptides have been identified in food proteins, mainly in milk, eggs, and plants. The industrial production of ACEI peptides is based on enzymatic proteolysis of whole food proteins, which leads to the release of small bioactive peptides with ACE-inhibitory activity. The problems associated to such procedures, namely, cost and loss of functional properties, have demonstrated the need to develop more straightforward methods to produce ACEI peptides. One viable hypothesis, discussed in this chapter, is to genetically engineer crop plants to produce and deliver antihypertensive peptides.

Keywords: plant engineering, recombinant protein, bioactive peptides, angiotensin I-converting enzyme, ACEI peptides

1. Introduction

The term "molecular pharming," blend of pharmaceutical and farming, surfaced in the literature in the 1980s to refer to the production of high-value compounds in transgenic animals. Nowadays, the expression is mainly employed to the production of recombinant pharmaceutically relevant proteins or secondary products in plants [1–3].

The roots of molecular pharming can be traced back to the mid-1980s when plants started to be genetically engineered to act as bioreactors that produced pharmaceutically relevant proteins. Barta et al. [4] demonstrated that tobacco and sunflower callus tissues were capable of expressing transcripts of a human growth hormone fusion gene. Although no protein was detected, this was the first report of plants expressing human genes and established plants as a potential production system for recombinant therapeutic proteins. Later on, the expression of a full-sized IgG in tobacco [5] was a major breakthrough since it revealed the ability of plants to produce complex functional mammalian proteins of pharmaceutical relevance. In 1990, the "authenticity" of plant-derived recombinant proteins was proved even further with the production of the first human protein (serum albumin) with confirmed native structure in tobacco and potato [6].

After several studies that demonstrated the capacity of various plant species and systems to produce recombinant pharmaceutical proteins and peptides, during the 1990s, the field of molecular pharming gained support and interest from the plant biotechnology community. The scientific attention was followed by commercial interest, with many start-up companies being created to capitalize the advantages of plants in relation to the established platforms. These advantages include being a more cost-effective, scalable, and safer means of producing pharmaceutically relevant proteins and peptides. In opposition to the fermentation-based traditional platforms that require a massive investment in bioreactors, plant-based production systems can be established with minimal investment and offer a myriad of different hosts and platforms [7]. However, the expectation that plants could easily compete for the market share of some well-established biopharmaceutical platforms, such as Chinese hamster ovary (CHO) cells, and that they could motivate the mainstream pharmaceutical industry to switch platforms was overinflated. The CHO epithelial cell lines are the most commonly used mammalian hosts for industrial production of therapeutic recombinant proteins. The technical limitations of plants, especially their lower yields compared to mammalian cell lines, allied to the colossal existing investment in fermentation infrastructures, the unfavorable public opinion on OGMs, and regulatory uncertainty, lead the mainstream pharmaceutical industry to be cautious and to a consequent stagnation of the molecular pharming field in the 2000s [8, 9]. This situation induced a change of paradigm concerning molecular pharming: the initial vision of a highly scalable and low-cost production system, while still valid, was replaced by the idea of a production system for certain niche products that are not easily manufactured by conventional systems [8, 9].

Molecular pharming embraces several platforms and technologies with different advantages and limitations, related by their use of plant tissues. Conversely to conventional biopharmaceutical production systems that are based on few selected platforms, particularly the bacterium *Escherichia coli*, yeasts such as *Pichia pastoris*, and mammalian cell lines such as Chinese hamster ovary (CHO) cells [3], pharming platforms range from plant cells or unicellular plants growing in bioreactors to whole plants growing in soil or hydroponic environments. Further, the technologies include stable integration of DNA into the nuclear genome or plastid genome and transient expression by infiltrating leaves with expression vectors based on *Agrobacterium tumefaciens*, plant viruses, or hybrids [3, 8]. This great diversity of molecular pharming confers adaptability and flexibility, allowing the selection of the most suitable platform for each product, but has also conduced to fragmentation. This fragmentation meant that in the early days of molecular pharming there was no driving force to establish molecular pharming as a single competitive platform. Consequently, no actions were made to match the industry requirements for high yields, standardized procedures, and good manufacturing practices (GMP) [7, 9]. More recently, efforts have been made to mimic the mainstream biopharmaceutical industry and place a focus only on a small number of platforms, namely, plant cell cultures, nuclear transgenic plants, and leafy plants transiently transformed [3, 10]. Since 2010 the attention of the biopharmaceutical industry to molecular pharming has been renewed as result of its consolidation on a small number of platforms and some target products that meet industry demands [8, 9].

In 2012, the FDA approval of the first recombinant plant-derived therapeutic for human use, Protalix Biotherapeutics' taliglucerase alfa (Elelyso™), was an important breakthrough for molecular pharming. The enzyme taliglucerase alfa is a carrot cell-expressed human recombinant β-glucocerebrosidase and is prescribed for the treatment of Gaucher's disease, a lysosomal storage disorder [11]. Imiglucerase, a recombinant form of glucocerebrosidase commercialized under the name Cerezyme®, was already produced in CHO cells. In this production platform, the enzyme required subsequent in vitro exposure to mannose residues in order to have biological activity, resulting in a time-consuming and expensive manufacturing process. Besides, this platform also has potential safety problems, namely, the risk of viral contamination, allergies, and other adverse reactions. In comparison, the plant-based platform is safer and less time-consuming and has reduced production costs, since the mannose units are posttranslationally added in vivo [11]. Glucocerebrosidase is a clear example of a target product in which safety, cost, and downstream processing issues were solved by switching from a traditional platform to molecular pharming. Another example that gathered mediatic exposure was ZMapp, a cocktail of three chimeric monoclonal antibodies targeting the Ebola virus surface glycoprotein produced in *Nicotiana benthamiana* using a hybrid transient expression system, the magnICON system. ZMapp was developed during the Ebola outbreak of 2014 by Mapp Biopharmaceutical Inc. (San Diego, USA), following initial studies on nonhuman primates [12]. ZMapp has since been used in humans under emergency compassionate protocols [13] and randomized controlled trials [14].

Following these examples of success, there has been a continuous increase in clinical trial applications and manufacturing capacity, which has also been correlated with the conception of more tangible regulations concerning plant-derived pharmaceuticals.

Although plants are still unlikely to substitute the established platforms [8], the recent promising developments in the field of molecular pharming demonstrate that glucocerebrosidase was not a lone case of success and that plant-based platforms could provide countless opportunities for the biopharmaceutical market. Plants combine the advantage of a full posttranslational modification potential with simple growth requirements and theoretically unlimited scalability in the case of field-grown whole plants. Plant-based platforms are versatile and allow the targeting of recombinant proteins and peptides produced to different organs or subcellular compartments, which provides an additional protection against proteolysis. Finally, plants are a safe host for therapeutic protein and peptide production since they do not harbor human or animal pathogens [15]. Therefore, instead of facing the red ocean of established pharmaceutical industries [16], molecular pharming is now evolving as a disruptive technology that creates its own marketplace by offering rapid drug development and production, unparalleled scalability, unique quality attributes such as tailored glycan structures, individualized therapies, and oral or topical applications of minimally processed plant tissues, thus reducing downstream costs [17].

2. Plant platforms for the production of therapeutic proteins and peptides

The continuous development of genetic engineering technologies for plants has resulted in an expansion of well-established plant-based platforms [18]. Molecular pharming encompasses platforms based on stably transformed whole-plants transgene insertion in the nuclear or plastid genome, transient expression using agroinfiltration, viral and hybrid vectors; microalgae and aquatic plants (e.g., duck-weed) stably transformed; and in vitro culture systems (e.g., cell suspensions, hairy roots, and moss protonema) [19]. Each platform has particular advantages and limitations; therefore its selection is done on a case-by-case basis, depending on economic considerations as well as on the product characteristics and intended use [20].

2.1 Platforms based on transgenic plants

Transgenic plants have been the most widely used platforms for recombinant protein production. To obtain stable transgenic lines, the gene encoding the desired protein is cloned into an expression construct, which generally includes a promoter and regulatory elements that ensure efficient RNA processing and protein synthesis [21]. This expression construct is then stably integrated into the plant nuclear genome, resulting in the stable inheritance of the transgene and expression of stable pharmaceutical proteins over generations [22]. Two major transformation strategies have been employed to insert expression constructs into the nuclear genome: *Agrobacterium*-mediated transformation in dicotyledonous species (dicots) and particle bombardment of DNA-coated gold or tungsten beads in monocotyledonous species (monocots) [3]. Transgenic plant lines offer several advantages as platforms for molecular pharming: they are suitable for long-term production of recombinant pharmaceutical proteins and are highly scalable, as each line can be used to produce seeds, which increase the number of plants in every generation. Ultimately, the production capacity of recombinant pharmaceutical proteins in transgenic plants is practically unlimited, as it only depends on the number of hectares available for the plant culture. The major drawbacks of transgenic plants are the long development and scale-up timescales, the unreliable production yields, and the potential spread of pharmaceutical crops in the environment and into the food chain by outcrossing and seed dispersal [3].

The development of simple transformation technologies has expanded the number of host plants available for molecular pharming. Currently, the major molecular pharming transgenic platforms are based on leafy crops, seeds, fruits, and vegetable crops. Leafy crops are benefic in terms of biomass yield and high soluble protein levels. Additionally, leaf harvesting does not need flowering and thus considerably reduces contamination through pollen or seed dispersal [23]. One disadvantage of leafy crops is that proteins are synthesized in an aqueous environment, which is more prone to protein degradation, resulting in lower production yields [24]. In fact, the mature leaves possess very large extra cytoplasmic vacuolar compartments containing various active proteolytic enzymes that are involved in the degradation of native and foreign proteins. This is particularly problematic in the case of therapeutic peptide production because short heterologous peptides have an inherent instability in plant cells [25]. In addition to the protein instability, the harvested material has limited shelf life and needs to be processed immediately after harvest.

Tobacco has been the most widely used leafy crop for molecular pharming. The major advantages of using tobacco to express pharmaceutical proteins are its high biomass yield, well-established technology for gene transfer and expression, year-round growth and harvesting, and the existence of large-scale infrastructure for processing [23]. However, the natural production of nicotine and other alkaloids in tobacco poses some safety issues in its use as a host system for heterologous protein production. Therefore, tobacco varieties with low nicotine and alkaloid levels have been produced to diminish the toxicity and overcome those safety issues. Recent studies have led to the approval of the first monoclonal antibody produced in transgenic tobacco plants, in phase I clinical trial [26]. Additionally, a 2018 publication reported the stable expression of adalimumab (a monoclonal antibody against *tumor necrosis factor-alpha (*TNF-α)) in tobacco plants [27]. Other leafy crops commonly used in molecular pharming include alfalfa and clover [19].

As an alternative to leafy crops, plant seeds have proven to be versatile hosts for recombinant proteins of all types, including peptides or short and long polypeptides as well as complex, noncontiguous proteins like antibodies and other immunoglobulins [28]. The expression of proteins in seeds can overcome the shortcomings of leafy crops in terms of protein stability and storage. Seeds possess specialized storage compartments, such as protein bodies and vacuoles, which provide the appropriate biochemical environment for protein accumulation, thus protecting the proteins expressed in seeds from proteolytic degradation [29]. Reports have demonstrated that antibodies expressed in seeds remain stable for at least 3 years at room temperature without detectable loss of activity [30]. Furthermore, the small size of most seeds permits to achieve a high recombinant protein concentration in a small volume, which facilitates extraction and downstream processing and reduces the costs of the overall manufacturing process [31]. One essential property of seeds is dormancy, which not only permits the stability of recombinant proteins but also allows a complete decoupling of the cycle of cultivation from the processing and purification of the protein [28]. Finally, proteins expressed in the seed do not normally interfere with vegetative plant growth, and this strategy also reduces exposure to herbivores and other nontarget organisms such as microbes in the biosphere [21]. Several crops have been studied for seed-based production, including cereals, such as maize, rice, barley, and wheat; legumes, such as pea and soybean; and oilseeds such as safflower and rapeseed. Maize has several advantages for seed-based expression of proteins; it has the highest biomass yield among food crops, and it is easy to transform, in vitro manipulate, and scale up [24]. These potentialities were explored by Prodigene Inc. for the production of the first commercially available plant-made protein, avidin (a protein with affinity for biotin used in biochemical assays). Other maize-derived protein products developed by this company include β-glucuronidase, aprotinin, laccase, and trypsin [32]. Prodigene was the first company to demonstrate the commercial benefits of plant-based platforms and was also a forerunner in the study of the economic impact of downstream processing in molecular pharming, having developed several successful approaches to recover intact and functional recombinant seeds from maize [3].

Maize has also been used to produce recombinant pharmaceutical proteins, including enzymes, vaccines, and antibodies [32, 33]. One of the most notable therapeutic proteins produced in maize is Meristem Therapeutics' gastric lipase, an enzyme intended for the treatment of exocrine pancreatic insufficiency—a disease significantly affecting cystic fibrosis sufferers—that has completed phase II clinical trial. In addition to this enzyme, Meristem Therapeutics has developed two other maize-derived products, human lactoferrin (phase I clinical trial), whose intellectual property was later acquired by Ventria Bioscience (http://www.ventria.com/), and collagen (pre-clinical stage).

Rice is another leading platform for recombinant protein and peptide production. Similar to maize, rice is easy to transform and scale up, but unlike maize, rice is self-pollinating, which reduces the risk of horizontal gene flow. Ventria Bioscience, in its ExpressTec platform, has used rice to produce recombinant pharmaceutical proteins, including human albumin, transferrin, lactoferrin, lysozyme, and vaccines against human rabies and Lyme disease. Its lead therapeutic candidate VEN100, whose active ingredient is lactoferrin, has been shown to reduce significantly antibiotic-associated diarrhea in high-risk patients and recently completed phase

II clinical trial [34]. Rice has also been widely used as host for peptide expression, especially for the production of allergen peptides (e.g., pollen and mite allergies) [35, 36]. Recent studies report that rice has the potential to offer an oral delivery system for vaccine antigens and therapeutic proteins and peptides [25, 35, 37].

Barley seeds have also been developed as commercial platforms. In comparison to other cereal crops, barley is less widely grown. However, this fact added to the self-pollinating nature of barley can be viewed as an advantage since the risk of contamination and outcrossing with non-transgenic crops is minimized. Considering this benefit, an Iceland-based company, ORF Genetics (https://orfgenetics.com/), has targeted barley grain as the expression host for several human cytokines and growth factors [19]. Other molecular pharming companies, such as Ventria Bioscience and Maltagen, have also been developing barley-based production platforms. Although barley is still recognized for its recalcitrance to transformation, over the last decade some progress has been made in the development of reliable transformation procedures [38].

The use of legume seeds, such as soybean and pea, for the production of recombinant pharmaceutical proteins, has been less explored than cereal-based platforms, with platforms based on legume seeds having yet to achieve commercial success. However, the fact that legume seeds have exceptionally high protein content (20–40%) can be exploited to achieve high yields of recombinant protein [39]. Soybean seeds have been used to express recombinant growth factors [40, 41], coagulation factors [42], and vaccine peptides [43]. Transgenic pea seeds have been previously used to produce a single-chain Fv fragment (scFV) antibody used in cancer diagnosis and therapy [44]. In another study, pea seeds were used to produce a vaccine that showed high immunogenicity and protection against rabbit hemorrhagic disease virus [45].

Safflower and rapeseed seeds are rich in oil and are, thus, referred as oilseeds. Oilseeds can provide useful recombinant pharmaceutical protein production systems. SemBioSys (http://www.sembiosys.ca/), with its oleosin-fusion platform, has been a pioneer in that field. Oleosins are the principal membrane proteins of oil bodies; oleosins confer peculiar structural properties to the oil bodies that offer simple extraction and purification procedures [46]. In the oleosin-fusion platform the recombinant protein is fused with oleosin and consequently targeted to the oil bodies. The fusion protein is then recovered through simple purification of the oil bodies and separated from oleosin by endoprotease digestion. Commercial production of hirudin in safflower by SemBioSys constituted the first report of an oilseed-derived protein [47]. The company has been focusing on safflower as its primary host ever since, with safflower-derived insulin being in phase I clinical trial [32].

Finally, fruit and vegetable crops can also be employed for molecular pharming. A major advantage of protein expression in fruit and vegetable crops is that edible organs can be consumed uncooked, unprocessed, or partially processed, making them particularly suitable for the production of recombinant subunit vaccines, nutraceuticals, and antibodies designed for topical application [29]. The oral delivery of recombinant therapeutics is one of the differentiating factor of molecular pharming in comparison to mainstream biopharmaceutical production systems, with several pharmaceutical products being produced in tomato fruits, potato tubers, and lettuce leaves for this purpose [3]. Tomato fruits are particularly useful for protein expression because the fruits are palatable as raw tissue but can also be lyophilized and stored for a long time [25]. Recently, human coagulation factor IX (hFIX) was expressed specifically in tomato fruits, constituting the first report on the expression of hFIX in plant [48]. Another study described the expression in tomato fruits of a thymosin α1 concatemer [49], an immune booster that plays an important role in the maturation, differentiation, and function of T cells.

The thymosin α1 concatemer derived from transgenic tomatoes exhibited biological activity and was proven to stimulate the proliferation of mice splenic lymphocytes in vitro. Moreover, thymosin α1 specific activity was higher when produced in tomato than in *Escherichia coli*, demonstrating the authenticity of the plant-made product. Other examples of tomato fruit expression include F1-V [50], a candidate subunit vaccine against pneumonic and bubonic plague, and β-secretase [51], to serve as a vaccine antigen against Alzheimer's disease.

In conclusion, platforms based on transgenic plants are a promising alternative to the conventional biopharmaceutical production platforms since they provide a stable source of pharmaceutical proteins and are also the most scalable of all molecular pharming platforms. This scalability of transgenic plants ensures the production of recombinant pharmaceutical proteins at levels previously inaccessible, namely, the commodity bulk production of monoclonal antibodies. In the current scenario of growing pharmaceutical demand, especially in developing countries, the use of transgenic plants can be game changing since they provide a highly scalable and low-cost means of producing medicines.

2.2 Platforms based on transplastomic plants

Transplastomic plants are a valuable alternative to transgenic plants for the production of recombinant pharmaceutical proteins. Transplastomic plants are obtained by the insertion of expression constructs into the plastid genome by particle bombardment. Since the *Agrobacterium* T-DNA (transfer DNA) complex is targeted to the nucleus, it is unsuitable for gene transfer to chloroplasts [24, 52]. Following the transformation procedure, the bombarded leaf explants are regenerated, and transplastomic plants with homoplastomic transformation (in which every chloroplast carries the transgene) are finally selected, recurring to a selection medium containing spectinomycin or in combination with streptomycin [53].

Plastid transformation can result in high yields of heterologous proteins because multiple copies of the genome are present in each plastid, and photosynthetic cells may contain hundreds or thousands of plastids [54]. As an example, the expression of a proteinaceous antibiotic in tobacco chloroplasts has achieved up to 70% of the total soluble proteins, which is the highest recombinant protein accumulation accomplished so far in plants [55]. Furthermore, chloroplasts provide a natural biocontainment of transgene flow since genes in chloroplast genomes are maternally inherited and consequently not transmitted through pollen, thereby avoiding unwanted escape into the environment. Other advantages of chloroplast engineering include the ability to express several genes as operons, and the accumulation of recombinant proteins in the chloroplast, thus reducing toxicity to the host plant [24].

Finally, transplastomic production platforms offer the possibility of oral delivery [54, 56]. In fact, it has been demonstrated that chloroplast-derived therapeutic proteins, delivered orally via plant cells, are protected from degradation in the stomach, probably due to the bioencapsulation of the therapeutic protein by the plant cell wall. They are subsequently released into the gut lumen by microbes that digest the plant cell wall, where the large mucosal intestine area offers an ideal system for oral drug delivery [57].

A shortcoming of expressing proteins via the chloroplast genome is that routine plastid engineering is still limited to tobacco, a crop that is not edible and thus unsuitable for oral delivery of therapeutic proteins. In addition, the synthesis of glycoproteins is not possible in the chloroplast system, as plastids do not carry out glycosylation [24]. Nevertheless, the expression of human somatotropin [58] in tobacco established that chloroplasts are capable of properly folding human proteins with disulfide bonds. In another study, the production of native cholera toxin B subunit

[59] demonstrated the capacity of chloroplasts to fold and assemble oligomeric proteins correctly. Other therapeutic proteins expressed in tobacco chloroplasts include interferons alpha-2a and alpha-2b [60, 61] and anti-cancer therapeutic agents such as human soluble tumor necrosis factor (TNF) [62] and azurin [63]. Recently, chloroplast transformation of lettuce has also been developed [64, 65] to provide oral delivery transplastomic systems [66, 67]. Several therapeutic proteins were produced in lettuce chloroplast, namely, proinsulin [66, 67], tuberculosis vaccine antigens [68], and human thioredoxin 1 protein [69]. The chloroplast production platform has yet to achieve commercial success, though the referred developments in this field augur a promising future for therapeutic protein production in chloroplasts.

2.3 Transient expression platforms

Transient expression is a phenomenon that occurs when genes are introduced into plant tissues and are expressed for a short period without stable DNA integration into the genome [3]. Traditionally, transient expression was used to verify expression construct activity and to test recombinant protein stability. This strategy allowed the identification and elimination of initial transformation problems, and thus the prospect of regenerating the desired transgenic lines was significantly improved. Recently, there has been an emergence of transient expression for the commercial production of recombinant pharmaceutical proteins. The advantages of transient expression platforms include the ease of manipulation, speed, low cost, and high yield of proteins. In comparison to transgenic plants, transient expression permits to achieve higher recombinant protein yields because there are no position effects (suppression of transgene expression by the surrounding genomic DNA following integration) [70].

Transient expression systems utilize the beneficial properties of plant pathogens to infect plants, spread systemically, and express transgenes at high levels, causing the rapid accumulation of recombinant proteins [8]. Currently, the major transient expression platforms are based on *Agrobacterium tumefaciens*, plant viruses, or hybrid vectors that utilize components of both (magnICON® technology).

The agroinfiltration method involves the vacuum infiltration of a suspension of recombinant *A. tumefaciens* into the plant leaf tissue, with the transgenes being then expressed from the uninterrupted T-DNA [8, 71]. Using this method, milligram amounts of recombinant protein are produced within a few weeks without the need to select transgenic plants, a process that takes months to years to be completed. This system has been commercially developed in tobacco [72] and alfalfa [73] but is also applicable to other crops such as lettuce [74], potato [75], and *Arabidopsis* [76]. An advantage of *Agrobacterium*-mediated transient expression is the fact that it allows to produce in plants complex proteins assembled from subunits [70].

Another transient expression technology is based on the use of plant viruses. In this technology, the gene of interest is inserted among viral replicating elements, episomically amplified and subsequently translated in the plant cell cytosol [77]. To date, the most efficient and high-yielding platforms have been developed using RNA viruses [78]. These plant viruses include *Tobacco mosaic virus* (TMV), *potato virus X* (PVX), and *Cowpea mosaic virus* (CPMV) (reviewed in [8]). The advantages of virus-based production include the rapid recombinant protein expression, the systemic spread of the virus, and the fact that multimeric proteins such as antibodies can also be produced by coinfecting plants with noncompeting vectors derived from different viruses [79, 80]. Transient expression vectors based on virus have been used to express peptides and long polypeptides (at least 140 amino acids long) as fusions to the coat protein, resulting in the assembly of chimeric virus particles (CVPs) displaying multiple copies of the peptide or polypeptide on its surface [77, 81].

Transient expression based in plant viruses has been commercially adopted by the now-closed Large Scale Biology Corporation (Vacaville, USA) that used a TMV-based vector for the production of patient-specific idiotype vaccines for the treatment of B-cell non-Hodgkin's lymphoma, which had successfully passed the phase I clinical trials [82].

Finally, the third transient expression strategy is based on hybrid systems that incorporate components of the T-DNA transfer and virus replication systems [3]. These hybrid systems use deconstructed viruses obtained by removing the coat protein (responsible for systemic movement) of the noncompeting virus strains and use *Agrobacterium* as the vehicle for the systemic delivery of the resulting viral vectors to the entire plant. These systems effectively address most of the major shortcomings of earlier plant-based technologies by providing the overall best combination of the following features: high expression level, high relative yield, low up- and downstream costs, very fast and low-cost R&D, and low biosafety concerns [83]. Consequently, there has been a commercial development based on several hybrid systems. One of most notable examples is the magnICON® system developed by Icon Genetics (https://www.icongenetics.com/) (formerly owned by Bayer Innovation, Dusseldorf, Germany; now a subsidiary of Nomad Bioscience, Halle, Germany), which features a deconstructed *Tobacco mosaic virus* (TMV) genome and *A. tumefaciens* as a delivery vehicle [83]. Another example is the iBioLaunch platform developed by the Fraunhofer Center for Molecular Biotechnology, which also features a deconstructed TMV genome [3]. Finally, the CPMV-HT platform is based on a deleted version of *Cowpea mosaic virus* RNA-2 and also allows the "hypertranslation" of recombinant proteins without virus spreading [8].

Examples of therapeutic recombinant proteins produced in these platforms have been generally reviewed in [3]. Recombinant protein production using transient expression is now being mobilized to a large scale with several companies developing scalable, automated plant-based GMP biomanufacturing facilities to efficiently produce large amounts of pharmaceuticals within weeks. Such facilities include the ones of the Fraunhofer Center for Molecular Biotechnology (Newark, DE) (https://www.fraunhofer.org/), Medicago Inc. (Quebec, Canada) (http://www.medicago.com/), Icon Genetics (Bayer; Halle, Germany) (http://www.icongenetics.com/), Texas A & M (College Station, TX), and Kentucky BioProcessing LLC (Owensboro, KY) (http://www.kbpllc.com/) [19].

In conclusion, the ability of transient plant expression systems to produce large quantities of recombinant protein, coupled to the use of current technology to increase yields, and the many promising technical solutions seems to be favorable compared with mammalian- or insect cell-based systems in quality, cost, and scale [19]. In case of emerging threats, transient platforms are advantageous since they produce large amounts of recombinant proteins rapidly (milligram quantities per plant within a few days) and can be scaled up quickly, currently providing the only reliable platform for rapid response situations [9]. During the H1N1 pandemic, the first batches of H1N1 virus-like particles (VLPs) could be produced by Medicago Inc. as soon as 3 weeks after the Centers for Disease Control and Prevention released the new influenza hemagglutinin sequence [73]. Similar lead times were reported for the H5N1 VLP vaccine [84]. Recently, the application of tobacco plant-based transient production systems, at Kentucky BioProcessing (KBP), to produce antibody lots against Ebola, was shown to significantly decrease the amount of time required for production over traditional methods, increase the quantity of antibody produced, and reduce the cost of manufacturing. Finally, at the other end of the market scale, transient expression platforms are economical for the production of pharmaceuticals for very small markets, such as orphan diseases and individualized therapies.

2.4 Callus and plant cell suspension cultures

Plant cell suspension cultures grow as individual cells or small aggregates and are usually derived from *callus* tissue by the disaggregation of friable callus pieces in shake bottles and are later scaled up for bioreactor-based production. Recombinant pharmaceutical protein production is achieved using transgenic explants to derive the cultures or by transforming the cells after disaggregation, usually by co-cultivation with *A. tumefaciens*. The co-cultivation of plant cell suspensions and recombinant *A. tumefaciens* has also been used for the transient expression of proteins [85]. Since these plant cell suspension cultures are grown in sterile contained environments, they provide a cGMP-compatible production environment that is more acceptable to the established pharmaceutical industry and regulatory authorities [3, 86]. These systems have added benefits of complex protein processing compared to bacteria and yeasts and increased safety compared to mammalian cell systems, which can harbor human pathogens. Another advantage of plant suspension cultures is the very low maintenance cost in comparison to other fermenter-based eukaryotic systems such as mammalian or insect cells. Moreover, the possible secretion of the target protein into the culture medium simplifies downstream processing and purification procedures [87, 88]. Nevertheless, plant cell cultures also have some limitations such as poor growth rates, somaclonal variation (particularly due to chromosomal rearrangements, common in plant cell cultures generated by *calli*), and gene silencing, together with the inhibition of product formation at high cell densities, formation of aggregates, cell wall growth, as well as shear-sensitivity for some species [89]. However, high levels of functional recombinant protein in plant cell suspension cultures were already obtained [87]. Besides, the previously mentioned first licensed recombinant pharmaceutical protein, Elelyso™, was produced in plant cell suspension cultures (reviewed in [88]). Tobacco has been the most popular source of suspension cells for recombinant protein production. Tobacco plants proliferate rapidly and are easy to transform, but other plant species have also been used to generate suspension cells, including rice and *Arabidopsis thaliana,* alfalfa, soybean, tomato, *Medicago truncatula*, and carrot [85, 88, 90]. Carrot suspension cells have been used by the aforementioned Protalix Biotherapeutics to produce a recombinant glucocerebrosidase. This case of commercial success shows that suspension cell cultures have potential as a viable system for large-scale protein production. Recently, carrot callus cultures, expressing epitopes from the cholesteryl ester transfer protein, were accessed for the potential of becoming an atherosclerosis oral vaccine [91].

3. Optimization of plant expression levels

The lower expression levels in comparison to the established biopharmaceutical platforms were one of the major obstacles for the commercialization of molecular pharming [9]. Therefore, numerous techniques have been developed to enhance protein expression, including codon optimization of protein sequences, to match the preferences of the host plant, targeting subcellular compartments that allow proteins to accumulate in a stable form; the use of strong, tissue-specific promoters; and the testing of different plant species and systems [25].

Protein synthesis can be increased by optimizing the components of the expression construct to maximize transcription, mRNA stability, and translation or by diminishing the impact of epigenetic phenomena that inhibit gene expression [92]. In this field, the general strategy is to use strong and constitutive promoters, such as the cauliflower mosaic virus 35S RNA promoter (CaMV 35S) and maize ubiquitin-1 promoter (ubi-1), for dicots and monocots, respectively. However, organ- and tissue-specific promoters are also being used to drive expression of the transgenes to a specific tissue or organ such as the tuber, the seed, and the fruit. Additionally, inducible promoters, whose activities are regulated by either chemical or external stimulus, may equally be used to prevent the lethality problem. Furthermore, transcription factors can also be used as boosters for the promoters to further enhance the expression level of the transgenes [53].

Protein stability can be increased by targeting proteins to cell compartments that reduce degradation. Protein targeting also affects the glycan structures added to proteins and the type of extraction and purification steps required to isolate the protein from the plant matrix. Proteins can be targeted to the secretory pathway by an N-terminal signal peptide, which is cleaved off for the release of the protein into the endoplasmic reticulum (ER). Proteins that do not require posttranslational modification, e.g., glycosylation, for their activity, can be targeted to the chloroplast using N-terminal transit peptides [93]. In addition, the target gene can be used to transform chloroplast directly, with highly enhanced protein accumulation. Moreover, posttranslational modifications of the ER lumen can also be avoided by expressing the protein as translational fusion with oleosin protein, which target the expression of the foreign protein to oil bodies of the seeds [28]. Other subcellular compartments like the protein-storing vacuoles are now being explored for recombinant protein accumulation, as it has been observed in rice seed endosperm [94].

4. Downstream processing

In the early years of molecular pharming, scientific studies were focused on demonstrating that plants could produce adequate quantities of recombinant pharmaceutical proteins and confer an oral delivery means. This led to downstream processing and the costs associated to it being basically overlooked. Downstream processing is now known to be an economically critical part of biomanufacturing processes (it can account for up to 80% of the total cost in a therapeutic protein production line) and also to be a key component of the regulatory process for evaluating the safety of pharmaceutical products [7]. The goal and the general steps for downstream processing are similar between plant and other expression systems: to recover the maximal amount of highly purified target protein with the minimal number of steps and at the lowest cost. The basic steps for downstream processes include tissue harvesting, protein extraction, purification, and formulation [22]. However, since in molecular pharming the costs of downstream processing are product-specific rather than platform-specific, the evaluation of downstream processing strategies and costs associated to it has to be done on a case-by-case basis. Nevertheless, even if unit operations have to be developed based on the properties of the product, others have to be developed based on the properties of the expression host. Plants produce process-related contaminants that require specific processing steps to ensure removal of fibers, oils, superabundant plant proteins such as RuBisCO, and potentially toxic metabolites such as the alkaloid nicotine in tobacco [8]. These secondary metabolites can be recovered from plant cells or tissues using methods such as adsorption, precipitation, and chromatography, often requiring phase partitioning and the use of mixtures of organic solvents. Several approaches have been used to facilitate downstream processing, including secretion of recombinant proteins, eliminating the plant cell disruption step; targeting of proteins into the protein bodies, oil bodies, or plastoglobules; and the use of affinity tags such as poly-histidine tags with the target protein, allowing protein purification by affinity chromatography [25]. In addition, oral delivery of whole plants or crude extracts containing the pharmaceutical relevant proteins can also be a way to simplify downstream processing and to easily distribute medicines to those in need. Furthermore, the optimization of plant's expression level can also ease downstream processing, with higher protein concentrations conducting to higher protein volumes [7].

Finally, several purification strategies have been investigated to separate target transgenic proteins from host plant proteins, which are tailored for each individual protein based on its solubility, size, pI, charge, hydrophobicity, or affinity to specific ligands, and the parallel characteristics of plant host proteins. Chromatographic methods, such as affinity chromatography, have been the most extensively used. However, recently increasing attention is being paid to non-chromatographic methods to provide alternatives for large-scale production [22].

5. Heterologous production of bioactive angiotensin I-converting enzyme inhibitory (ACEI) peptides

In the broad range of known bioactive peptides, angiotensin I-converting enzyme inhibitory (ACEI) peptides derived from food proteins have attracted particular attention and have been studied the most comprehensively for their ability to prevent hypertension [95]. In this chapter we will further focus on the possibility to genetically engineer crop plants to produce and deliver antihypertensive ACEI peptides, therefore creating alternative sources to fight hypertension and prevent cardiovascular disease.

5.1 Cardiovascular disease and the renin-angiotensin system

Cardiovascular disease (CVD) has been recognized as the leading cause of death in developed countries. Hypertension or high blood pressure is one of the major independent risk factors for CVD [96]. States of CVD include conditions such as coronary heart disease, peripheral artery disease, and stroke. Hypertension is a condition defined by a blood pressure measurement of 140/90 mmHg or above and is thought to affect up to 30% of the worldwide adult population [95]. The kinin-nitric oxide (KNO) system and the renin-angiotensin system (RAS), **Figure 1**, play a crucial role in the control of hypertension by the action of angiotensin I-converting key enzyme (EC 3.4.15.1; ACE) [96–99].

Several synthetic ACE inhibitors such as captopril, enalapril, and lisinopril have been prescribed for the treatment of hypertension, congestive heart failure, and diabetic neuropathy [100]. However, their consumption is associated with various side effects including cough, skin rashes, hypotension, loss of taste, angioedema, reduced renal function, and fetal abnormalities [95]. The side effects associated to synthetic ACE inhibitors and the high prevalence of hypertension have led scientists to search for natural and safer therapies. Interestingly, the study of ACEI peptides has revealed that they do not have significant effects on blood pressure in normotensive subjects, suggesting a convenient mechanism that avoids acute hypotensive effects. Based on this finding, it is hypothesized that ACEI peptides could be used in initial treatment of mildly hypertensive individuals or even as supplemental treatments [101].

5.2 Antihypertensive ACEI peptides

So far, several ACEI peptides have been identified in food proteins, mainly in milk, eggs, and plants, currently constituting the most well-known class of bioactive peptides [102–104]. These peptides are inactive within the sequence of parent proteins, but they can be released by enzymatic proteolysis in vivo or

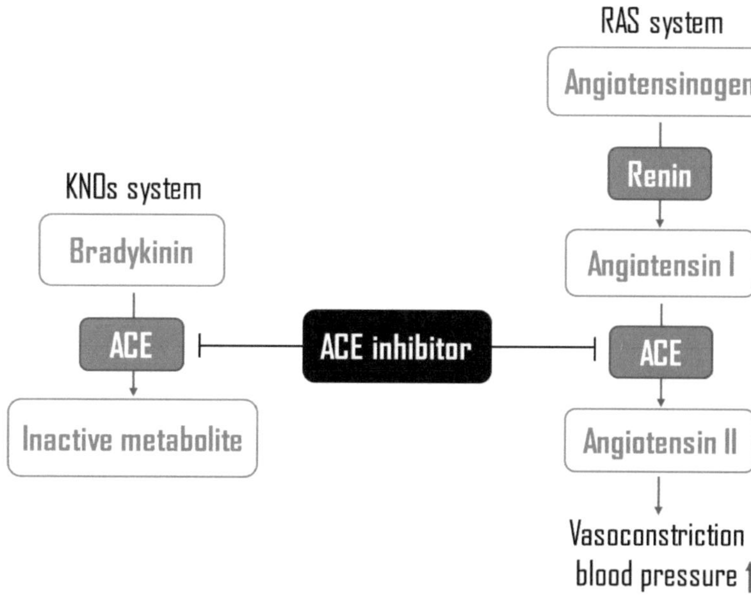

Figure 1.
The kinin-nitric oxide (KNO) system and the renin-angiotensin system (RAS). The left side (KNO system) shows the mechanism of the action of ACEI on ACE that cleaves bradykinin, a nonapeptide acting as vasodilatory hormone, and causes the formation of an inactive heptapeptide. In the right side (RAS system), the inhibition of ACE activity plays an important physiological role in regulation of blood pressure by inhibiting the conversion of the hormone angiotensin I to angiotensin II, a potent vasoconstrictor (figure adapted from Erdmann et al. [96]).

in vitro, for example, during gastrointestinal digestion or during food processing. A common feature shared by the majority of ACEI peptides is the generally short sequence, i.e., 2–12 amino acids in length. However, some larger inhibitory sequences have been identified in milk fermented with *Enterococcus faecalis* [105] and *Lactobacillus casei* Shirota [106], in koumiss [107], tuna [108], bonito [109], and rotifer [110]. Studies have also indicated that binding to ACE is strongly influenced by the substrate's C-terminal tripeptide sequence. Hydrophobic amino acid residues with aromatic or branched side chains at each of the C-terminal tripeptide positions are common features among potent inhibitors. The presence of hydrophobic Pro residues at one or more positions in the C-terminal tripeptide region seems to positively influence a peptide's ACE-inhibitory activity [95]. In general, the peptides showing higher activity against ACE have Tyr, Phe, Trp, or Pro at their C-terminus [95]. The peptides TQVY from rice [111], MRW from spinach [112], and YKYY from wakame [113] are some examples of this principle. **Table 1** reviews some examples of ACEI activities of plant origin, whose peptides responsible for such activity may be potential sources for the heterologous production of ACEI peptides.

The most common method to produce and identify ACEI peptides is through enzymatic hydrolysis of food proteins with gastrointestinal enzymes such as pepsin and trypsin or with commercial proteases such as Alcalase™ [127]. ACEI peptides have also been produced with *Lactobacillus, Lactococcus lactis*, and *E. faecalis* strains during milk fermentation [105, 106]. Nevertheless, there are problems associated to this type of industrial production of ACEI peptides, including the difficulty to isolate the peptide of interest from the complex mixture of compounds produced by enzymatic hydrolysis, the high cost, low recovery, and the low bioavailability. These disadvantages denote the need to develop new and alternative approaches for their production.

Source	ACEI activity (IC50; μM)	Antihypertensive activity (mmHg)	Dose (mg/kg)	Reference
Chlorella vulgaris	29.6	Not determined	—	[114]
Chebulic myrobalan	100	Not determined	—	[115]
Bitter melon	8.64	−31.5 to −36.3	2–10	[116]
Mung bean	13.4	Not determined	—	[117]
Pea	64	Not determined	—	[118]
Peanut	72	Not determined	—	[100]
Potato	18–86*	Not determined	—	[119]
Rapeseed	28	−11.3	7.5	[120]
Rice	18.2	−40	30	[111]
Soybean	14–39*	−17.5	2	[121]
Soybean	21	Not determined	—	[122]
Soybean	1.69	Not determined	—	[123]
Soybean	17.2	Not determined	—	[124]
Spinach	0.6–4.2*	−13.5 to −20*	20–100	[112]
Wakame	21–213*	−50	50	[113]
Walnut	25.7	Not determined	—	[125]
Wheat	20	Not determined	—	[126]

**Different values for the same plant product related to the ACEI peptide sequence.*

Table 1.
Examples of ACEI peptide activity from different plant origin.

5.3 Heterologous production of ACEI peptides in plants

In recent years, the application of recombinant DNA technologies for the production of ACEI peptides at a large scale and low cost has gathered attention in the biotechnology community. Investigation has been focused on the development of expression methods for antihypertensive peptide production in different plant crops [128]; and here, we tried to provide some promising examples.

Thus far, the main strategies that have been adopted are as follows: the over-expression of ACEI peptide precursor proteins and the production of particular peptides as heterologous components [101], the modification of some storage proteins to produce chimeric proteins carrying ACEI peptides [101], and also the generation of multimer proteins containing tandem repeats of ACEI peptides, flanked by protease recognition sequences that allow the peptide release during gastrointestinal digestion.

5.3.1 Rice

Transgenic rice plants that accumulate novokinin (RPLKPW), a potent antihypertensive peptide designed according to the structure of ovokinin (2–7) (RADHPF), as a fusion with the rice storage protein glutelin, have been generated. The engineered peptide is expressed under the control of endosperm-specific glutelin promoters and specifically accumulates in seeds. Oral administration of either the RPLKPW-glutelin fraction or transgenic rice seeds to spontaneously hypertensive rats (SHRs)—the main model for assessing the in vivo activity of ACEI peptides (e.g., [108, 111, 122])—significantly reduced systolic blood pressures, suggesting the possible application of transgenic rice seed as a nutraceutical delivery system and particularly for administration of antihypertensive peptides [129].

Wakasa et al. [130] attempted the generation of transgenic rice seeds that would accumulate higher amounts of novokinin peptide by expressing 10 or 18 tandemly repeated novokinin sequences, with the KDEL endoplasmic reticulum retention signal at the C-terminus, and using the glutelin promoter along with its signal peptide. Although the chimeric protein was unexpectedly located in the nucleolus and the accumulation was low, a significant antihypertensive activity was detected after a single oral dose to SHRs. More importantly, this effect was observed over a relatively longer duration time, with intervals of 5 weeks between doses as low as 0.0625 g transgenic seeds per kg.

5.3.2 Soybean

Soybean [*Glycine max* (L.) Merr.] is an attractive option for the production of ACEI peptides given that soybean seeds contain a large amount of total protein. Therefore, there has been an effort to generate soybean lines with improved ACEI properties foreseeing the creation of novel functional foods.

Matoba et al. [128], introduced novokinin (RPLKPW) into homologous sequences of a soybean β-conglycinin α' subunit by site-directed mutagenesis. Founded on first achievements from an *E. coli* expressed protein, the muted β-conglycinin α' subunit carrying novokinin repeats were also expressed in soybean. This chimeric protein accumulated at levels of up to 0.2% of extracted protein from transgenic soybean seeds [131]. Still, the levels of expression were too low, and it was not possible to assess the in vivo effects of these soybean seeds.

Novokinin has also been expressed in transgenic soybean seeds in a fusion form along with a β-conglycinin α' subunit. Interestingly, a reduced systolic blood pressure was observed in SHRs after administering a dose of 0.15 g kg^{-1} of protein extracts. A similar effect was attained following administration of a 0.25 g kg^{-1} dose of defatted flour. Thus, it was concluded that this chimeric protein produced in soybean possessed an antihypertensive activity [132].

Additionally, a synthetic gene of His-His-Leu (HHL), an ACEI peptide derived from a Korean soybean paste, was tandemly multimerized to a 40-mer, ligated with ubiquitin as a fusion gene (UH40), and subsequently expressed in *E. coli*. Following digestion with leucine aminopeptidase, the 405-Da HHL monomer was recovered by reverse-phase high-performance liquid chromatography (HPLC). MALDITOF mass spectrometry, glutamine-TOF mass spectrometry, N-terminal sequencing, and measurement of ACE-inhibiting activity confirmed that the resulting peptide was the HHL [133]. The potential use of this antihypertensive chimeric protein in soybean has yet to be assessed.

5.3.3 Tomato and tobacco

A modified version of amarantin, the main seed storage protein of *Amaranthus hypochondriacus*, carrying four tandem repeats of the ACEI dipeptide Val-Tyr into the acidic subunit of amarantin, was expressed in cell suspension cultures of *Nicotiana tabacum* L. NT1. Protein hydrolysates obtained from transgenic *calli* showed high levels of inhibition of the angiotensin-converting enzyme, with an IC50 value of 3.5 μg ml^{-1}, and 10-fold lower levels than that of protein extracts of wild-type cells (IC50 of 29.0 μg ml^{-1}) [134]. This was the first time that a chimeric protein comprising an ACEI peptide was produced in plant cell suspension cultures.

This modified version of amarantin was also expressed in the fruit of transgenic tomato plants. Protein hydrolysates from transgenic tomato fruits showed in vitro ACE inhibition, with IC50 values ranging from 0.376 to 3.241 μg ml^{-1};

this represented an increase of up to 13-fold in the inhibitory activity when compared with the protein hydrolysates of non-transformed fruits [135]. These two results suggest the possible application of tobacco plant cell suspension cultures and transgenic tomato fruits for massive production of this engineered version of amarantin, which could be especially used as an alternative hypertension therapy [134, 135].

5.3.4 Amaranth

Although amaranth has not been genetically modified to produce ACEI peptides, the feasibility of developing a modified amarantin acidic subunit has been widely assessed [129, 134–139]. Recently, the in vivo effect of an *E. coli*-modified amarantin protein, four units of Val-Tyr dipeptides (VY) in tandem, and one of Ile-Pro-Pro tripeptides (IPP) incorporated in the amarantin acidic subunit (AMC3) was evaluated in SHRs in a one-time oral administration experiment. This study showed that enzymatic hydrolysates of AMC3-containing ACEI peptide (4xVY and IPP) sequences had significant in vivo antihypertensive action [138]. The positive reports of amarantin expression in *E. coli* [136, 138, 139] along with the sustained expression of amarantin-modified proteins in tobacco [134] and tomato [135] prospect the successful production of ACEI peptide fusion proteins in amaranth.

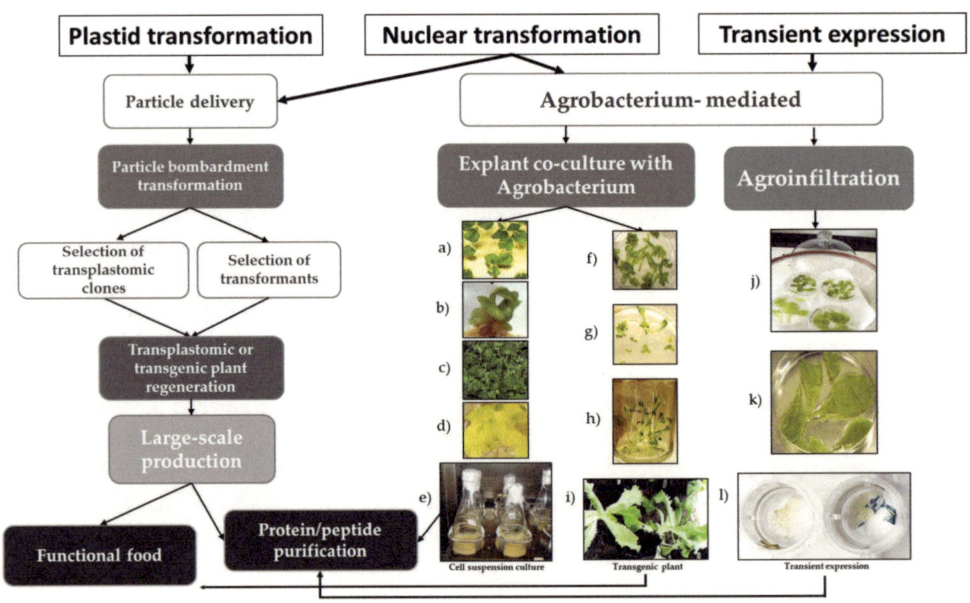

Figure 2.
Schematic representation of the technologies involved in different plant platforms for the production of therapeutically important proteins and peptides. ***Plastid transformation*** *by particle bombardment can result in regeneration of transplastomic plants, revealing high-yield heterologous production, with the possibility of protein/peptide oral delivery or purification.* ***Nuclear transformation*** *can be accomplished by particle bombardment or by Agrobacterium-mediated transformation, resulting in the regeneration of stable transgenic plants. Finally, the technology based on* ***transient expression,*** *here with the example of agroinfiltration. We present Medicago truncatula and lettuce as examples: (a) M. truncatula co-culture of leaf explants with Agrobacterium, (b) and (c) plant regeneration via somatic embryogenesis according to Araújo et al. [145], (d) and (e) establishment of a cell suspension culture from callus for protein/peptide production [146, 147], (f) lettuce leaf explant co-culture with Agrobacterium, (g) and (h) plant regeneration via shoot organogenesis at PCB lab, (i) lettuce transgenic plants which can be used for oral delivery, (j) and (k) agroinfiltration of lettuce leaf explants according to Negrouk et al. [74], and (l) example of a control explant (left) and transient expression of a 35S::GUS(int) cassette in lettuce leaves (right).*

5.3.5 Lettuce and Medicago truncatula

Lettuce (*Lactuca sativa*) is a commercially important crop belonging to the Asteraceae family. It is a diploid (2n = 18), autogamous species with a genome size of 2.7 Gb [140]. This crop is particularly suitable for oral delivery of therapeutics as its raw leaves are consumed by humans, and the time to obtain an edible product is only weeks, compared to the months needed for crops such as tomato or potato. Therefore, recently lettuce has been investigated as a production host for edible recombinant therapeutics [66, 67, 141]. Furthermore, the fact that stable transformation procedures for both nuclear [142] and plastid genomes [64], and transient expression [74], are widely available, is also an advantage. Lettuce has been used as production host for several recombinant therapeutics, virus-like particles (VLPs) and monoclonal antibodies [143], antigens [142, 144], and human therapeutic proteins [66, 69].

Medicago truncatula is a model plant from the legume family. It is a diploid (2n = 16), autogamous species, with a relatively small genome and short life cycle of 3–5 months. These characteristics enable this species to be used in molecular genetic studies and expression of foreign genes [145]. The phylogenetic distance to economically important crops is crucial in the choice of this plant by many researchers and funding agencies, since it allows comparative studies within the legume family. The methodologies for the establishment of long-term cell suspension culture are well recognized [146], and the potential of *M. truncatula* as expression host has also been established for the production of feed additives [20, 87], human hormones [90], and human enzymes [147].

The use of these two species in molecular pharming is at the center of a recent collaboration between the Plant Cell Biotechnology (PCB) Laboratory (ITQB UNL), the Cell Differentiation and Regeneration Laboratory (iBiMED UA), and the Institute of Plant Genetics (IPG PAS). This cooperation foresees the usage of these two species as exceptional hosts for the heterologous production and/or delivery of ACEI peptides, and a resume of this ongoing project is here schematically presented (**Figure 2**). This figure also provides an overview of the technologies involved in different plant platforms discussed in this chapter.

6. Conclusions

Molecular pharming has been recently and extensively reviewed, and the future of this technology has gathered some optimistic expectations. A myriad of studies have already demonstrated the capacity of various plant species and systems to produce recombinant pharmaceutical proteins and peptides. This technology has already been put to the test in case of emerging threats, where transient platforms proved to be strategic for rapid production of large amounts of recombinant proteins in response to pandemic situations. However, their usefulness for the production of functional foods still falls short of expectations, as well as the attainment of its full potential in bioactive peptide production. With the improvement of known plant platforms and development of new genetic engineering techniques and their exploration, it is forthcoming an evolution in the production of heterologous bioactive peptides, to which we hope to contribute with our ACEI pharming project. The advent of genome editing techniques (with the advantage of site- specific gene insertion), like the CRISPR/Cas9 methodology, will undoubtedly increase and democratize plant transformation events and will certainly contribute to the increase of genetically modified species for molecular pharming purposes.

Acknowledgements

We acknowledge financial support from Fundação para a Ciência e Tecnologia (FCT), Portugal, through the research Grant SFRH/BPD/74784/2010 (Duque AS) and funding, through the research unit GREEN-it "Bioresources for Sustainability" (UID/Multi/04551/2013) and FCT/COMPETE/QREN/EU-FEDER (for Institute of Biomedicine iBiMED: UID/BIM/ 04501/2013). We also acknowledge the Program for Scientific Cooperation FCT/Poland 2017/2018 and Angelini Pharmaceuticals and PREMIVALOR (for Angelini University Award 2012/2013).

Author details

Carolina Gomes[1], Filipe Oliveira[2,3], Sandra Isabel Vieira[2†] and Ana Sofia Duque[3*†]

1 Institute of Plant Genetics, Polish Academy of Sciences (IPG PAS), Poznań, Poland

2 Cell Differentiation and Regeneration Laboratory, Department of Medical Sciences, Institute of Biomedicine (iBiMED), Universidade de Aveiro, Aveiro, Portugal

3 Plant Cell Biotechnology Laboratory, Green-it Unit, Instituto de Tecnologia Química e Biológica António Xavier (ITQB NOVA), Oeiras, Portugal

*Address all correspondence to: sduque@itqb.unl.pt

† Equally contributing authors

References

[1] Lössl AG, Clarke JL. Conference scene: Molecular pharming: Manufacturing medicines in plants. Immunotherapy. 2013;**5**:9-12

[2] Ma JKC, Christou P, Chikwamba R, et al. Realising the value of plant molecular pharming to benefit the poor in developing countries and emerging economies. Plant Biotechnology Journal. 2013;**11**:1029-1033

[3] Paul MJ, Teh AYH, Twyman RM, et al. Target product selection—Where can molecular pharming make the difference? Current Pharmaceutical Design. 2013;**19**:5478-5485

[4] Barta A, Sommergruber K, Thompson D, et al. The expression of a nopaline synthase-human growth hormone chimaeric gene in transformed tobacco and sunflower callus tissue. Plant Molecular Biology. 1986;**6**:347-357

[5] Hiatt A, Caffferkey R, Bowdish K. Production of antibodies in transgenic plants. Nature. 1989;**342**:76-78

[6] Sijmons PC, Dekker BM, Schrammeijer B, et al. Production of correctly processed human serum albumin in transgenic plants. Nature Biotechnology. 1990;**8**:217-221

[7] Fischer R, Schillberg S, Hellwig S, et al. GMP issues for recombinant plant-derived pharmaceutical proteins. Biotechnology Advances. 2012;**30**:434-439

[8] Stöger E, Fischer R, Moloney M, et al. Plant molecular pharming for the treatment of chronic and infectious diseases. Annual Review of Plant Biology. 2014;**65**:743-768

[9] Fischer R, Schillberg S, Buyel JF, et al. Commercial aspects of pharmaceutical protein production in plants. Current Pharmaceutical Design. 2013;**19**:5471-5477

[10] Fischer R, Vasilev N, Twyman RM, et al. High-value products from plants: The challenges of process optimization. Current Opinion in Biotechnology. 2015;**32**:156-162

[11] Zimran A, Brill-Almon E, Chertkoff R, et al. Pivotal trial with plant cell-expressed recombinant glucocerebrosidase, taliglucerase alfa, a novel enzyme replacement therapy for Gaucher disease. Blood. 2011;**118**:5767-5773

[12] Qiu X, Wong G, Audet J, et al. Reversion of advanced Ebola virus disease in nonhuman primates with ZMapp. Nature. 2014;**514**:47-53

[13] Lyon GM, Mehta AK, Varkey JB, et al. Clinical care of two patients with Ebola virus disease in the United States. The New England Journal of Medicine. 2014;**371**:2402-2409

[14] Davey RT, Dodd L, Proschan MA, et al. A randomized, controlled trial of ZMapp for Ebola virus infection. The New England Journal of Medicine. 2016;**375**:1448-1456

[15] Karg SR, Kallio PT. The production of biopharmaceuticals in plant systems. Biotechnology Advances. 2009;**27**:879-894

[16] Kim WC, Mauborgne R. Blue Ocean Strategy: How to Create Uncontested Market Space and Make the Competition Irrelevant. Boston, Massachusetts: Harvard Business School Press; 2005

[17] Sabalza M, Christou P, Capell T. Recombinant plant-derived pharmaceutical proteins: Current technical and economic bottlenecks. Biotechnology Letters. 2014;**36**:2367-2379

[18] Paul M, JK-C M. Plant-made pharmaceuticals: Leading products and production platforms. Biotechnology and Applied Biochemistry. 2011;**58**:58-67

[19] Xu J, Dolan MC, Medrano G, et al. Green factory: Plants as bioproduction platforms for recombinant proteins. Biotechnology Advances. 2012;**30**:1171-1184

[20] Abranches R, Marcel S, Arcalis E, et al. Plants as bioreactors: A comparative study suggests that *Medicago truncatula* is a promising production system. Journal of Biotechnology. 2005;**120**:121-134

[21] Commandeur U, Twyman RM, Fischer R. The biosafety of molecular farming in plants. AgBiotechNet. 2003;**5**:1-9

[22] Chen Q. Expression and purification of pharmaceutical proteins in plants. Biological Engineering. 2008;**1**:291-321

[23] Makhzoum A, Benyammi R, Moustafa K, et al. Recent advances on host plants and expression cassettes' structure and function in plant molecular pharming. BioDrugs. 2014;**28**:145-159

[24] Ma JKC, Drake PMW, Christou P. The production of recombinant pharmaceutical proteins in plants. Nature Reviews. Genetics. 2003;**4**:794-805

[25] Lico C, Santi L, Twyman RM, et al. The use of plants for the production of therapeutic human peptides. Plant Cell Reports. 2012;**31**:439-451

[26] Ma JK-C, Drossard J, Lewis D, et al. Regulatory approval and a first-in-human phase I clinical trial of a monoclonal antibody produced in transgenic tobacco plants. Plant Biotechnology Journal. 2015;**13**:1106-1120

[27] Zvirin T, Magrisso L, Yaari A, et al. Stable expression of adalimumab in *Nicotiana tabacum*. Molecular Biotechnology. 2018;**60**:387-395

[28] Boothe J, Nykiforuk C, Shen Y, et al. Seed-based expression systems for plant molecular farming. Plant Biotechnology Journal. 2010;**8**:588-606

[29] Twyman RM, Stoger E, Schillberg S, et al. Molecular farming in plants: Host systems and expression technology. Trends in Biotechnology. 2003;**21**:570-578

[30] Stöger E, Vaquero C, Torres E, et al. Cereal crops as viable production and storage systems for pharmaceutical scFv antibodies. Plant Molecular Biology. 2000;**42**:583-590

[31] Stöger E, Ma JK-C, Fischer R, et al. Sowing the seeds of success: Pharmaceutical proteins from plants. Current Opinion in Biotechnology. 2005;**16**:167-173

[32] Lau OS, Sun SSM. Plant seeds as bioreactors for recombinant protein production. Biotechnology Advances. 2009;**27**:1015-1022

[33] Hayden CA, Fischer ME, Andrews BL, et al. Oral delivery of wafers made from HBsAg-expressing maize germ induces long-term immunological systemic and mucosal responses. Vaccine. 2015;**33**:2881-2886

[34] Laffan AM, McKenzie R, Forti J, et al. Lactoferrin for the prevention of post-antibiotic diarrhoea. Journal of Health, Population, and Nutrition. 2011;**29**:547-551

[35] Takagi H, Hiroi T, Hirose S, et al. Rice seed ER-derived protein body as an efficient delivery vehicle for oral tolerogenic peptides. Peptides. 2010;**31**:1421-1425

[36] Suzuki K, Kaminuma O, Yang L, et al. Prevention of allergic asthma by

vaccination with transgenic rice seed expressing mite allergen: Induction of allergen-specific oral tolerance without bystander suppression. Plant Biotechnology Journal. 2011;**9**:982-990

[37] Takagi H, Hiroi T, Yang L, et al. A rice-based edible vaccine expressing multiple T cell epitopes induces oral tolerance for inhibition of Th2-mediated IgE responses. Proceedings of the National Academy of Sciences. 2005;**102**:17525-17530

[38] Mrízová K, Holasková E, Öz MT, et al. Transgenic barley: A prospective tool for biotechnology and agriculture. Biotechnology Advances. 2014;**32**:137-157

[39] Vianna GR, Cunha NB, Murad AM, et al. Soybeans as bioreactors for biopharmaceuticals and industrial proteins. Genetics and Molecular Research. 2011;**10**:1733-1752

[40] Ding S-H, Huang L-Y, Wang Y-D, et al. High-level expression of basic fibroblast growth factor in transgenic soybean seeds and characterization of its biological activity. Biotechnology Letters. 2006;**28**:869-875

[41] Cunha NB, Murad AM, Cipriano TM, et al. Expression of functional recombinant human growth hormone in transgenic soybean seeds. Transgenic Research. 2011;**20**:811-826

[42] Cunha NB, Murad AM, Ramos GL, et al. Accumulation of functional recombinant human coagulation factor IX in transgenic soybean seeds. Transgenic Research. 2011;**20**:841-855

[43] Maruyama N, Fujiwara K, Yokoyama K, et al. Stable accumulation of seed storage proteins containing vaccine peptides in transgenic soybean seeds. Journal of Bioscience and Bioengineering. 2014;**118**:441-447

[44] Perrin Y, Vaquero C, Gerrard I, et al. Transgenic pea seeds as bioreactors for the production of a single-chain Fv fragment (scFV) antibody used in cancer diagnosis and therapy. Molecular Breeding. 2000;**6**:345-352

[45] Mikschofsky H, Schirrmeier H, Keil GM, et al. Pea-derived vaccines demonstrate high immunogenicity and protection in rabbits against rabbit haemorrhagic disease virus. Plant Biotechnology Journal. 2009;**7**:537-549

[46] Bhatla SC, Kaushik V, Yadav MK. Use of oil bodies and oleosins in recombinant protein production and other biotechnological applications. Biotechnology Advances. 2010;**28**:293-300

[47] Parmenter DL, Boothe JG, van Rooijen GJH, et al. Production of biologically active hirudin in plant seeds using oleosin partitioning. Plant Molecular Biology. 1995;**29**:1167-1180

[48] Zhang H, Zhao L, Chen Y, et al. Expression of human coagulation factor IX in transgenic tomato (*Lycopersicon esculentum*). Biotechnology and Applied Biochemistry. 2007;**48**:101-107

[49] Chen Y, Wang A, Zhao L, et al. Expression of thymosin α1 concatemer in transgenic tomato (*Solanum lycopersicum*) fruits. Biotechnology and Applied Biochemistry. 2009;**52**:303

[50] Alvarez ML, Pinyerd HL, Crisantes JD, et al. Plant-made subunit vaccine against pneumonic and bubonic plague is orally immunogenic in mice. Vaccine. 2006;**24**:2477-2490

[51] Kim H-S, Youm J-W, Moon K-B, et al. Expression analysis of human β-secretase in transgenic tomato fruits. Protein Expression and Purification. 2012;**82**:125-131

[52] Gelvin SB. *Agrobacterium*-mediated plant transformation: The biology behind the "gene-jockeying" tool. Microbiology and Molecular Biology Reviews. 2003;**67**:16-23

[53] Obembe OO, Popoola JO, Leelavathi S, et al. Advances in plant molecular farming. Biotechnology Advances. 2011;**29**:210-222

[54] Daniell H, Singh ND, Mason H, et al. Plant-made vaccine antigens and biopharmaceuticals. Trends in Plant Science. 2009;**14**:669-679

[55] Oey M, Lohse M, Kreikemeyer B, et al. Exhaustion of the chloroplast protein synthesis capacity by massive expression of a highly stable protein antibiotic. The Plant Journal. 2009;**57**:436-445

[56] Rubio-Infante N, Govea-Alonso DO, Alpuche-Solís ÁG, et al. A chloroplast-derived C4V3 polypeptide from the human immunodeficiency virus (HIV) is orally immunogenic in mice. Plant Molecular Biology. 2012;**78**:337-349

[57] Kwon K-C, Daniell H. Oral delivery of protein drugs bioencapsulated in plant cells. Molecular therapy: The Journal of the American Society of Gene Therapy. 2016;**24**:1342-1350

[58] Staub JM, Garcia B, Graves J, et al. High-yield production of a human therapeutic protein in tobacco chloroplasts. Nature Biotechnology. 2000;**18**:333-338

[59] Daniell H, Lee S-B, Panchal T, et al. Expression of the native cholera toxin B subunit gene and assembly as functional oligomers in transgenic tobacco chloroplasts. Journal of Molecular Biology. 2001;**311**:1001-1009

[60] Arlen PA, Falconer R, Cherukumilli S, et al. Field production and functional evaluation of chloroplast-derived interferon-alpha2b. Plant Biotechnology Journal. 2007;**5**:511-525

[61] Nurjis F, Khan M. Expression of recombinant interferon α-2a in tobacco chloroplasts using micro projectile bombardment. African Journal of Biotechnology. 2002;**10**:17016-17022

[62] Wang D, Bai X, Bai X, et al. Expression of human soluble tumor necrosis factor (TNF)-related apoptosis-inducing ligand in transplastomic tobacco. African Journal of Biotechnology. 2002;**10**:6816-6823

[63] Roh KH, Choi SB, Kwak B-K, et al. A single cupredoxin azurin production in transplastomic tobacco. Plant Biotechnol Rep. 2014;**8**:421-429

[64] Lelivelt CLC, McCabe MS, Newell CA, et al. Stable plastid transformation in lettuce (*Lactuca sativa* L.). Plant Molecular Biology. 2005;**58**:763-774

[65] Kanamoto H, Yamashita A, Asao H, et al. Efficient and stable transformation of *Lactuca sativa* L. cv. Cisco (lettuce) plastids. Transgenic Research. 2006;**15**:205-217

[66] Ruhlman T, Ahangari R, Devine A, et al. Expression of cholera toxin B-proinsulin fusion protein in lettuce and tobacco chloroplasts-oral administration protects against development of insulitis in non-obese diabetic mice. Plant Biotechnology Journal. 2007;**5**:495-510

[67] Boyhan D, Daniell H. Low-cost production of proinsulin in tobacco and lettuce chloroplasts for injectable or oral delivery of functional insulin and C-peptide. Plant Biotechnology Journal. 2011;**9**:585-598

[68] Lakshmi PS, Verma D, Yang X, et al. Low cost tuberculosis vaccine antigens in capsules: Expression in chloroplasts, bio-encapsulation, stability and functional evaluation in vitro. PLoS One. 2013;**8**:e54708

[69] Lim S, Ashida H, Watanabe R, et al. Production of biologically active human thioredoxin 1 protein in lettuce chloroplasts. Plant Molecular Biology. 2011;**76**:335-344

[70] Komarova TV, Baschieri S, Donini M, et al. Transient expression systems

for plant-derived biopharmaceuticals. Expert Review of Vaccines. 2010;**9**:859-876

[71] Kapila J, De Rycke R, Van Montagu M, et al. An *Agrobacterium*-mediated transient gene expression system for intact leaves. Plant Science. 1997;**122**:101-108

[72] Whaley KJ, Hiatt A, Zeitlin L. Emerging antibody products and *Nicotiana* manufacturing. Human Vaccines. 2011;**7**:349-356

[73] D'Aoust M-A, Couture MM-J, Charland N, et al. The production of hemagglutinin-based virus-like particles in plants: A rapid, efficient and safe response to pandemic influenza. Plant Biotechnology Journal. 2010;**8**:607-619

[74] Negrouk V, Eisner G, Lee H, et al. Highly efficient transient expression of functional recombinant antibodies in lettuce. Plant Science. 2005;**169**:433-438

[75] Bhaskar PB, Venkateshwaran M, Wu L, et al. *Agrobacterium*-mediated transient gene expression and silencing: A rapid tool for functional gene assay in potato. PLoS One. 2009;**4**:1-8

[76] Kim MJ, Baek K, Park C-M. Optimization of conditions for transient *Agrobacterium*-mediated gene expression assays in *Arabidopsis*. Plant Cell Reports. 2009;**28**:1159-1167

[77] Lico C, Chen Q, Santi L. Viral vectors for production of recombinant proteins in plants. Journal of Cellular Physiology. 2008;**216**:366-377

[78] Sainsbury F, Lomonossoff GP. Extremely high-level and rapid transient protein production in plants without the use of viral replication. Plant Physiology. 2008;**148**:1212-1218

[79] Verch T, Yusibov V, Koprowski H. Expression and assembly of a full-length monoclonal antibody in plants using a plant virus vector. Journal of Immunological Methods. 1998;**220**:69-75

[80] Giritch A, Marillonnet S, Engler C, et al. Rapid high-yield expression of full-size IgG antibodies in plants coinfected with noncompeting viral vectors. Proceedings of the National Academy of Sciences. 2006;**103**:14701-14706

[81] Gleba Y, Klimyuk V, Marillonnet S. Viral vectors for the expression of proteins in plants. Current Opinion in Biotechnology. 2007;**18**:134-141

[82] McCormick AA, Reddy S, Reinl SJ, et al. Plant-produced idiotype vaccines for the treatment of non-Hodgkin's lymphoma: Safety and immunogenicity in a phase I clinical study. Proceedings of the National Academy of Sciences. 2008;**105**:10131-10136

[83] Gleba Y, Klimyuk V, Marillonnet S. Magnifection-a new platform for expressing recombinant vaccines in plants. Vaccine. 2005;**23**:2042-2048

[84] Landry N, Ward BJ, Trépanier S, et al. Preclinical and clinical development of plant-made virus-like particle vaccine against avian H5N1 influenza. PLoS One. 2010;**5**(12):e15559

[85] Schillberg S, Raven N, Fischer R, et al. Molecular farming of pharmaceutical proteins using plant suspension cell and tissue cultures. Current Pharmaceutical Design. 2013;**19**:5531-5542

[86] Spök A, Twyman RM, Fischer R, et al. Evolution of a regulatory framework for pharmaceuticals derived from genetically modified plants. Trends in Biotechnology. 2008;**26**:506-517

[87] Pires AS, Cabral MG, Fevereiro P, et al. High levels of stable phytase accumulate in the culture medium of

transgenic *Medicago truncatula* cell suspension cultures. Biotechnology Journal. 2008;**3**:916-923

[88] Santos RB, Abranches R, Fischer R, et al. Putting the spotlight back on plant suspension cultures. Frontiers in Plant Science. 2016;7:297

[89] Hellwig S, Drossard J, Twyman RM, et al. Plant cell cultures for the production of recombinant proteins. Nature Biotechnology. 2004;**22**:1415-1422

[90] Pires AS, Rosa S, Castanheira S, et al. Expression of a recombinant human erythropoietin in suspension cell cultures of *Arabidopsis*, tobacco and Medicago. Plant Cell, Tissue and Organ Culture. 2012;**110**:171-181

[91] Govea-Alonso DO, Tello-Olea MA, Beltrán-López J, et al. Assessment of carrot callus as biofactories of an atherosclerosis oral vaccine prototype. Molecular Biotechnology. 2017;**59**:482-489

[92] Twyman RM, Schillberg S, Fischer R. Optimizing the yield of recombinant pharmaceutical proteins in plants. Current Pharmaceutical Design. 2013;**19**:5486-5494

[93] Kmiec B, Teixeira PF, Glaser E. Shredding the signal: Targeting peptide degradation in mitochondria and chloroplasts. Trends in Plant Science. 2014;**19**:771-778

[94] Ou J, Guo Z, Shi J, et al. Transgenic rice endosperm as a bioreactor for molecular pharming. Plant Cell Reports. 2014;**33**:585-594

[95] Roseanne N, FitzGerald R. Antihypertensive peptides from food proteins. Food & Function. 2013;**3**:350-361

[96] Erdmann K, Cheung BWY, Schröder H. The possible roles of food-derived bioactive peptides in reducing the risk of cardiovascular disease. The Journal of Nutritional Biochemistry. 2008;**19**:643-654

[97] Li G-H, Le G-W, Shi Y-H, et al. Angiotensin I-converting enzyme inhibitory peptides derived from food proteins and their physiological and pharmacological effects. Nutrition Research. 2004;**24**:469-486

[98] Murray BA, FitzGerald RJ. Angiotensin converting enzyme inhibitory peptides derived from food proteins: Biochemistry, bioactivity and production. Current Pharmaceutical Design. 2007;**13**:773-791

[99] Belović M, Mastilović J, Torbica A, et al. Potential of bioactive proteins and peptides for prevention and treatment of mass non-communicable diseases. Food and Feed Research. 2011;**38**:51-62

[100] Jimsheena VK, Gowda LR. Arachin derived peptides as selective angiotensin I-converting enzyme (ACE) inhibitors: Structure-activity relationship. Peptides. 2010;**31**:1165-1176

[101] Rosales-Mendoza S, Paz-Maldonado LMT, Govea-Alonso DO, et al. Engineering production of antihypertensive peptides in plants. Plant Cell, Tissue and Organ Culture. 2013;**112**:159-169

[102] Iwaniak A, Minkiewicz P, Darewicz M. Food-originating ACE inhibitors, including antihypertensive peptides, as preventive food components in blood pressure reduction. Comprehensive Reviews in Food Science and Food Safety. 2014;**13**:114-134

[103] Ibrahim HR, Ahmed AS, Miyata T. Novel angiotensin-converting enzyme inhibitory peptides from caseins and whey proteins of goat milk. Journal of Advanced Research. 2017;**8**:63-71

[104] Daskaya-Dikmen C, Yucetepe A, Karbancioglu-Guler F, et al. Angiotensin-I-converting enzyme

(ACE)-inhibitory peptides from plants. Nutrients. 2017;**9**:316

[105] Quirós A, Ramos M, Muguerza B, et al. Identification of novel antihypertensive peptides in milk fermented with *Enterococcus faecalis*. International Dairy Journal. 2007;**17**:33-41

[106] Rojas-Ronquillo R, Cruz-Guerrero A, Flores-Nájera A, et al. Antithrombotic and angiotensin-converting enzyme inhibitory properties of peptides released from bovine casein by *Lactobacillus casei* Shirota. International Dairy Journal. 2012;**26**:147-154

[107] Chen Y, Wang Z, Chen X, et al. Identification of angiotensin I-converting enzyme inhibitory peptides from koumiss, a traditional fermented mare's milk. Journal of Dairy Science. 2010;**93**:884-892

[108] Lee S-H, Qian Z-J, Kim S-K. A novel angiotensin I converting enzyme inhibitory peptide from tuna frame protein hydrolysate and its antihypertensive effect in spontaneously hypertensive rats. Food Chemistry. 2010;**118**:96-102

[109] Hasan F, Kumada Y, Hashimoto N, et al. Fragmentation of angiotensin-I converting enzyme inhibitory peptides from bonito meat under intestinal digestion conditions and their characterization. Food and Bioproducts Processing. 2006;**84**:135-138

[110] Lee JK, Hong S, Jeon J-K, et al. Purification and characterization of angiotensin I converting enzyme inhibitory peptides from the rotifer, *Brachionus rotundiformis*. Bioresource Technology. 2009;**100**:5255-5259

[111] Li G-H, Qu M-R, Wan J-Z, et al. Antihypertensive effect of rice protein hydrolysate with in vitro angiotensin I-converting enzyme inhibitory activity in spontaneously hypertensive rats. Asia Pacific Journal of Clinical Nutrition. 2007;**16**:275-280

[112] Yang Y, Marczak ED, Yokoo M, et al. Isolation and antihypertensive effect of angiotensin I-converting enzyme (ACE) inhibitory peptides from spinach Rubisco. Journal of Agricultural and Food Chemistry. 2003;**51**:4897-4902

[113] Suetsuna K, Nakano T. Identification of an antihypertensive peptide from peptic digest of wakame (*Undaria pinnatifida*). The Journal of Nutritional Biochemistry. 2000;**11**:450-454

[114] Sheih I-C, Fang TJ, Wu T-K. Isolation and characterisation of a novel angiotensin I-converting enzyme (ACE) inhibitory peptide from the algae protein waste. Food Chemistry. 2009;**115**:279-284

[115] Sornwatana T, Bangphoomi K, Roytrakul S, et al. Chebulin: *Terminalia chebula* Retz. fruit-derived peptide with angiotensin-I-converting enzyme inhibitory activity. Biotechnology and Applied Biochemistry. 2015;**62**:746-753

[116] Priyanto AD, Doerksen RJ, Chang C-I, et al. Screening, discovery, and characterization of angiotensin-I converting enzyme inhibitory peptides derived from proteolytic hydrolysate of bitter melon seed proteins. Journal of Proteomics. 2015;**128**:424-435

[117] Li G-H, Wan J-Z, Le G-W, et al. Novel angiotensin I-converting enzyme inhibitory peptides isolated from Alcalase hydrolysate of mung bean protein. Journal of Peptide Science. 2006;**12**:509-514

[118] Jakubczyk A, Karaś M, Baraniak B, et al. The impact of fermentation and in vitro digestion on formation angiotensin converting enzyme (ACE) inhibitory peptides from pea proteins. Food Chemistry. 2013;**141**:3774-3780

[119] Pihlanto A, Akkanen S, Korhonen HJ. ACE-inhibitory and antioxidant properties of potato (*Solanum tuberosum*). Food Chemistry. 2008;**109**:104-112

[120] Marczak ED, Usui H, Fujita H, et al. New antihypertensive peptides isolated from rapeseed. Peptides. 2003;**24**:791-798

[121] Chen J-R, Okada T, Muramoto K, et al. Identification of angiotensin I-converting enzyme inhibitory peptides derived from the peptic digest of soybean protein. Journal of Food Biochemistry. 2002;**26**:543-554

[122] Kodera T, Nio N. Identification of an angiotensin I-converting enzyme inhibitory peptides from protein hydrolysates by a soybean protease and the antihypertensive effects of hydrolysates in 4 spontaneously hypertensive model rats. Journal of Food Science. 2006;**71**:C164-C173

[123] Gouda KGM, Gowda LR, Rao AGA, et al. Angiotensin I-converting enzyme inhibitory peptide derived from glycinin, the 11S globulin of soybean (*Glycine max*). Journal of Agricultural and Food Chemistry. 2006;**54**:4568-4573

[124] Vallabha V, Tiku PK. Antihypertensive peptides derived from soy protein by fermentation. International Journal of Peptide Research and Therapeutics. 2014;**20**:161-168

[125] Liu M, Du M, Zhang Y, et al. Purification and identification of an ACE inhibitory peptide from walnut protein. Journal of Agricultural and Food Chemistry. 2013;**61**:4097-4100

[126] Thewissen BG, Pauly A, Celus I, et al. Inhibition of angiotensin I-converting enzyme by wheat gliadin hydrolysates. Food Chemistry. 2011;**127**:1653-1658

[127] Pihlanto A, Makine S. Antihypertensive properties of plant protein derived peptides. In: Bioactive Food Peptides in Health and Disease. Rijeka, Croatia: InTech; 2013. DOI: 10.5772/54565

[128] Matoba N, Doyama N, Yamada Y, et al. Design and production of genetically modified soybean protein with anti-hypertensive activity by incorporating potent analogue of ovokinin. FEBS Letters. 2001;**497**:50-54

[129] Yang L, Tada Y, Yamamoto MP, et al. A transgenic rice seed accumulating an anti-hypertensive peptide reduces the blood pressure of spontaneously hypertensive rats. FEBS Letters. 2006;**580**:3315-3320

[130] Wakasa Y, Zhao H, Hirose S, et al. Antihypertensive activity of transgenic rice seed containing an 18-repeat novokinin peptide localized in the nucleolus of endosperm cells. Plant Biotechnology Journal. 2011;**9**:729-735

[131] Nishizawa K, Kita A, Doi C, et al. Accumulation of the bioactive peptides, novokinin, LPYPR and rubiscolin, in seeds of genetically modified soybean. Bioscience, Biotechnology, and Biochemistry. 2008;**72**:3301-3305

[132] Yamada Y, Nishizawa K, Yokoo M, et al. Anti-hypertensive activity of genetically modified soybean seeds accumulating novokinin. Peptides. 2008;**29**:331-337

[133] Jeong D-W, Shin DS, Ahn C-W, et al. Expression of antihypertensive peptide, His-His-Leu, as tandem repeats in *Escherichia coli*. Journal of Microbiology and Biotechnology. 2007;**17**:952-959

[134] Santos-Ballardo DU, Germán-Báez LJ, Cruz-Mendívil A, et al. Expression of the acidic-subunit of amarantin, carrying the antihypertensive biopeptides VY, in cell suspension

cultures of *Nicotiana tabacum* NT1. Plant Cell, Tissue and Organ Culture. 2013;**113**:315-322

[135] Germán-Báez LJ, Cruz-Mendívil A, Medina-Godoy S, et al. Expression of an engineered acidic-subunit 11S globulin of amaranth carrying the antihypertensive peptides VY, in transgenic tomato fruits. Plant Cell, Tissue and Organ Culture. 2014;**118**:305-312

[136] Luna-Suárez S, Medina-Godoy S, Cruz-Hernández A, et al. Modification of the amaranth 11S globulin storage protein to produce an inhibitory peptide of the angiotensin I converting enzyme, and its expression in *Escherichia coli*. Journal of Biotechnology. 2010;**148**:240-247

[137] Castro-Martínez C, Luna-Suárez S, Paredes-López O. Overexpression of a modified protein from amaranth seed in *Escherichia coli* and effect of environmental conditions on the protein expression. Journal of Biotechnology. 2012;**158**:59-67

[138] Medina-Godoy S, Rodríguez-Yáñez SK, Bobadilla NA, et al. Antihypertensive activity of AMC3, an engineered 11S amaranth globulin expressed in *Escherichia coli*, in spontaneously hypertensive rats. Journal of Functional Foods. 2013;**5**:1441-1449

[139] Morales-Camacho JI, Dominguez-Dominguez J, Paredes-Lopez O. Overexpression of a modified amaranth protein in *Escherichia coli* with minimal media and lactose as inducer. Recent Patents on Biotechnology. 2013;**7**:61-70

[140] Truco MJ, Ashrafi H, Kozik A, et al. An ultra-high-density, transcript-based, genetic map of lettuce. G3: Genes, Genomes, Genetics. 2013;**3**:617-631

[141] Martínez-González L, Rosales-Mendoza S, Soria-Guerra RE, et al. Oral immunization with a lettuce-derived *Escherichia coli* heat-labile toxin B subunit induces neutralizing antibodies in mice. Plant Cell, Tissue and Organ Culture. 2011;**107**:441-449

[142] Liu C-W, Chen JJW, Kang C-C, et al. Transgenic lettuce (*Lactuca sativa* L.) expressing H1N1 influenza surface antigen (neuraminidase). Scientia Horticulturae. 2012;**139**:8-13

[143] Lai H, He J, Engle M, et al. Robust production of virus-like particles and monoclonal antibodies with geminiviral replicon vectors in lettuce. Plant Biotechnology Journal. 2012;**10**:95-104

[144] Huy N-X, Yang M-S, Kim T-G. Expression of a cholera toxin B subunit-neutralizing epitope of the porcine epidemic diarrhea virus fusion gene in transgenic lettuce (*Lactuca sativa* L.). Molecular Biotechnology. 2011;**48**:201-209

[145] Araújo SS, Amaral Duque ASRL, Dos Santos DMMF, et al. An efficient transformation method to regenerate a high number of transgenic plants using a new embryogenic line of *Medicago truncatula* cv. Jemalong. Plant Cell, Tissue and Organ Culture. 2004;**78**:123-131

[146] Duque AS, Pires AS, Dos SDM, et al. Efficient somatic embryogenesis and plant regeneration from long-term cell suspension cultures of *Medicago truncatula* cv. Jemalong. In Vitro Cellular & Developmental Biology-Plant. 2006;**42**:270-274

[147] Pires AS, Santos RB, Nogueira AC, et al. Production of human lipocalin-type prostaglandin D synthase in the model plant *Medicago truncatula*. In Vitro Cellular & Developmental Biology-Plant. 2014;**50**:276-281

Chapter 5

Polymeric Nanocomposite-Based Agriculture Delivery System: Emerging Technology for Agriculture

Mohammad Ashfaq, Neetu Talreja, Divya Chuahan and Werayut Srituravanich

Abstract

The increasing global population has forced the agricultural area to enhance the yield of crop, thereby fulfilling the requirements of people. The advancement has led to synthesis of nanomaterials with different size, shapes, and biocompatibility aspects towards specific applications like agriculture. Several nanomaterials such as metal, metal oxide, carbon nanotubes (CNTs), carbon nanofibers (CNFs), graphene, and its derivatives have shown potential ability for augmenting the yield of crops and protect crops against pathogens. However, these nanomaterials required smart delivery system that might easily deliver the nanofertilizers in a controlled manner. In this context, the incorporation of nanotechnology and polymer science might be developing newer technology with minimal usage and maximum effectiveness for improvement of crops. The incorporation of nanomaterials in polymeric composites offers newer approaches for agricultural delivery system that might provide various advantages such as higher stability, solubility, uniform distribution, and controlled release. Moreover, nanomaterials have potential ability for advancement in the genetic engineering. Herein, we discuss the role of nanomaterials in the growth of the plant, polymeric nanocomposite materials for agriculture delivery system with the advancement in the genetic engineering, and future prospects of these polymeric-nanocomposite materials in agriculture.

Keywords: nanomaterials, polymeric composite, nanofertilizers, delivery system

1. Introduction

Nanomaterials (NMs) have attracted great interest especially in the field of agriculture that enhanced productivity of crops with lesser cost and waste [1, 2]. NMs offer sustainable effectiveness in the field of agriculture including protection and production of crops [3]. The significant advancement and development of the newer agricultural technologies is sturdily required because of continuously increasing food requirements globally [4]. The global food production must be increased around 70–100% by 2050 to achieve the demand of growing population [5, 6]. In this context, the agriculture promoted from various innovative

IntechOpen

technologies such as hybrid species, synthesis chemicals, and biotechnological developments [7]. However, continuous production of agricultural crops might be one of the great challenges due to the lack of nutrients/changes in climates. To overcome such issues related with the loss of production or improvement in the yield of crops, farmers continuously used agrochemicals. Nonetheless, excessive use of these agrochemicals leads to deterioration of soil, degradation of agro-ecosystems, and environmental problems [8, 9]. In this context, NMs have a technological advancement, might be transformed and allied sectors that provides newer agricultural tools for the management of stresses (biotic and abiotic), detection of diseases, improved nutrients absorption ability, and translocation ability. On the other hand, NMs might help to understand agricultural biology as well as interaction of nanomaterials with plants, thereby enhancing the nutritional value as well as productivity of the crops. However, the exact role of NMs in agriculture still remains a concern.

Numerous NMs including carbon-based nanomaterials (single-walled carbon nanotubes (SW-CNTs), multi-walled carbon nanotubes (MW-CNTs) [10, 11], carbon nanofibers (CNFs), graphene and fullerenes [12–15], metal and its oxide-based nanomaterials [16–18], magnetized iron (Fe) nanoparticles [19], aluminum oxide (Al_2O_3) [20], copper (Cu) [21], gold (Au) [22, 23], silver (Ag) [24, 25], silica (Si) [26], zinc (Zn) nanoparticles and zinc oxide (ZnO) [27–29], titanium dioxide (TiO_2) [30], and cerium oxide (Ce_2O_3) [31], etc.) and bio-composite nanomaterials have been developed. These NMs are efficiently used in the field of agriculture for production and protection of crops [32–35]. However, phytotoxicity, degradation of soil, large-scale production, agglomeration, and effective delivery system still remain a concern. On the other hand, CNFs have the potential ability to deliver micronutrients in plants and the release of micronutrients (Cu/Zn nanoparticles) in a controlled manner. However, CNFs also required polymeric delivery system for real applications [34]. In this context, polymeric nanocomposite has emerged as one of the most promising tools for the delivery of micronutrients and agrochemicals in the plant system [36].

Several polymers such as polyvinyl alcohol (PVA), chitosan, polyvinyl-pyrrolidone (PVP), starch, hyaluronic acid (HA), poly(lactic-co-glycolic acid) (PLGA), poly-lactic acid (PLA), etc. have been used as a carrier for delivery system for various biological applications due to their high biocompatibility, biodegradability, nontoxicity, cost-effectiveness, and excellent film forming ability [37–39]. Various processes such as cross-linking, emulsion formation, and self-assembly have been used for the synthesis of polymeric nanocomposite that facilitate controlled release of agrochemical/micronutrients within the plants. The encapsulation of nanomaterials by using polymeric matrix also aided advantages to enhance effectiveness of the nanomaterials, decreasing cellular toxicity and environmental contaminations [40]. On the other hand, smart polymeric materials and delivery system have the potential ability to deliver the genes/biomolecules/micronutrients within the plants and also protect viruses and pathogens [41, 42]. This book chapter focuses on the various nanomaterials and polymeric composite that augment the plant growth and interaction of nanomaterials with plants, genes/biomolecules/micronutrient delivery and discuss the advancement of genetic engineering by using nanomaterials.

2. Emergence of engineering nanomaterials (ENMs)

The advancement has led to the synthesis of engineering nanomaterials (ENMs) of various sizes and shapes [43]. This advancement in the synthesis routes offers great interest to develop unique characteristics against specific end applications like production and protection of crops [18, 44, 45]. Interestingly, these ENMs have been used in various applications such as medicine, environmental science, and sensors. Nonetheless, nanomaterial use in agriculture mainly for improvement in crop yield and crop protection is an under-explored research area. On the other hand, preliminary studies suggested that nanomaterials incorporated with polymers or polymeric composite have the potential ability to improve germination of seeds and growth of the plants, protection of crops, detection of pathogens, and detection of pesticide. The synthetic polymers also play a crucial role in agriculture because of polymeric materials that are pH-sensitive [46], temperature-sensitive [47], and climatic responsive that might be beneficial for growth of the plants such as mulches, shelters, and greenhouses (for fumigation, irrigation, and controlling water distribution) [48]. The ideal polymers for agricultural applications should have various properties such as stability, transmission, permeability, and weather ability, which is one of the important concerns nowadays [49]. In this context, functionalization of the polymers or polymeric composites has received significant consideration for the production of newer polymeric composites with improved characteristics [50]. **Figure 1** shows schematic representation of nanomaterials and their agricultural applications.

Several polymeric nanomaterials like chitosan, PVA, lipids, and PLGA are used in agriculture for augmenting the growth and protection of plants. The uptake and efficiency of the nanomaterials vary with the species, discussed later in the text.

In general, reactive nanomaterials exhibit various end applications due to their active functional groups and characteristic ability of polymers. Therefore, ENMs might be successfully utilized in different end applications including agriculture.

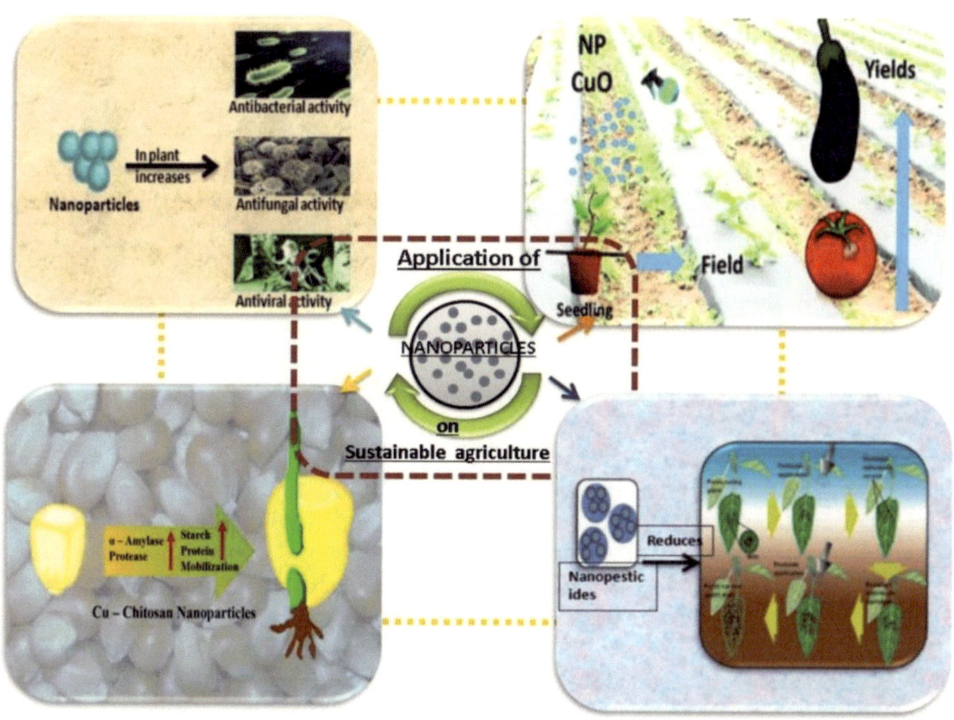

Figure 1.
Schematic representation of nanomaterials and its agricultural applications. Reprint permission Prasad et al. [7], copyright © 2017 Prasad, Bhattacharyya and Nguyen creative commons attribution license (CC BY).

3. Polymeric composites

Polymers are mainly used for the controlled release of agrochemicals such as insecticides, pesticides, fungicides, germicides, and growth stimulants. There are various factors such as cost, climate condition, controlled release, simple formulation, biocompatibility, and biodegradability involved in alteration in polymers for targeted system or applications. Moreover, thermal stability, thermal plasticity, glass-transition state, nature of polymers, melting point, its compatibility with biologically active molecules, and desired shape and size of the product still remain a concern. On the other hand, these polymers have the potential ability to control the release rate and rate of biodegradability, thereby being effective in various end applications, mainly medicine and agriculture [51]. The control release behavior of the polymeric formulation is one of the most important advantages in the delivery system like medicine, agrochemicals, and micronutrients.

Usually, the controlled release system is mainly divided into two groups; (1) encapsulation of active molecules/agrochemicals/micronutrients by using polymeric matrix and (2) polymeric matrix and active molecules/agrochemicals/micronutrients enclosed and formation of macromolecular backbones. Several polymers (natural, synthetic, and synthetic elastomers) such as carboxymethyl cellulose [52], cellulose acetate phthalate [53], gelatin [54], chitosan [55], gum Arabic [56], polylactic acid (PLA) [57], poly-butadiene, poly-lactic-glycolic acid (PLGA) [58], polyhydroxyalkanoates (PHAs) [59], polyvinyl alcohol (PVA) [60], polyacrylamide [61, 62], and polystyrene, etc., [63] are extensively used in various delivery systems. Among all of them, natural polymers are extensively used for controlled release of drugs/agrochemicals because of their low cost and being biodegradable. Moreover, controlled release rate might be tuned by using different molecular weight-based polymers and cross-linking of different polymers; therefore, polymeric composites are efficiently used in various biological applications. Recently, nanomaterials have been used as nanofertlizers, nanopesticides, and nanomaterials for genetic advancement, treatment of plant disease, and improved growth of the plants.

In general, polymers encapsulated with various materials including metal nanoparticles, carbon-based nanomaterials, biological molecules, agrochemicals, pesticides, insecticides, etc. with controlled release behaviors enhance the biocompatibility of the materials and are easy for applicability and thereby effectively used in various end applications, mainly medicine and agriculture [7].

3.1 Metal-polymer composites

The metal nanomaterials such as Cu, Zn, Fe, titanium dioxide (TiO_2), aluminum oxide (Al_2O_3), silicon dioxide (SiO_2), aluminum nitride (AlN), boron nitride (BN), and zinc oxide (ZnO), etc. are extensively used for the plant growth and protection of crops. Usually, nanomaterials are synthesized for providing the controlled release delivery system for agrochemicals that enhanced solubility and protecting biologically active molecules against early degradation, thereby enhancing effectiveness of agrochemicals even at lower doses. However, these metal-based nanomaterials accumulate on the root and translocate less within the shoot and leave. Moreover, agglomeration, instability, and difficulty to use directly in land still remain a concern. In this context, the continuously increasing demands of the hybrid materials provide newer technological breakthrough for various end applications such as medical, environment, sensors, and agriculture. There are various existing materials such as plastics, metals, ceramic, and polymers that cannot achieve technological requirements for different applications. Usually, hybrid nanomaterials containing various nanomaterials as a filler with the polymeric matrix have great interest because of various advantages like high biocompatibility, controlled release, stability, and nontoxicity. The most important dominating approach for synthesis of metal-polymer composite by using metal/metal oxide encapsulates with polymers produces the desired product. The enhancements of material characteristic by using filler-polymer interactions at the interface as well as the uniform dispersion of the nanomaterials within the polymeric matrix. Usually, there are three approaches to achieve these requirements: (1) alteration of fillers/nanomaterial properties, (2) alteration of polymer properties by functionalization or formation of co-polymers, and (3) developing desired properties with the hybrid materials/polymeric nanocomposite. On the basis of agricultural applications, polymeric coating or polymeric nanocomposite is important, as higher concentration of nanomaterials might cause some extent of toxicity within the plants [64, 65].

3.2 Carbon-polymer composites

Carbon-based nanomaterials exhibited various end applications such as environmental remediation, sensors, drug delivery, antibacterial agents, crop protection, and growth regulator of the plants due to unique characteristics, mainly optical, electrical, mechanical, and thermal properties. The significant development has been done in the synthesis of carbon-based nanomaterials such as activated carbon, activated carbon fibers, CNTs, CNFs, graphene, and fullerenes, which have great interest in agriculture due to their possibility as a growth stimulant and protection of crops [66–79]. Moreover, these carbon-based nanomaterials, mainly CNTs, and CNFs, have the potential ability to penetrate seed coat as well as translocation ability within the plants from root to shoot to leaves. Several reports suggested that CNTs and CNFs efficiently translocate within the plants. These studies suggested that carbon-based nanomaterials acted as a growth stimulant with increasing the uptake of water and nutrients. Interestingly, CNFs hold metal nanoparticles and the release of metal nanoparticles in a control manner. These metal nanoparticles like Cu, Zn, and Fe also acted as micronutrients for the plants; therefore CNFs acted as a carrier for micronutrient delivery. Moreover, CNFs increase the water uptake ability, germination rate, and nontoxicity even at higher concentration of dose and therefore are used as a growth stimulant of the plants. In a recent study, Gupta et al. suggested that the CNFs are used as a carrier to deliver acylated-homoserine lactone in chick pea plants [34]. The study suggested that CNF-acylated homoserine lactone-based composite increased the plant growth as well as stress tolerance ability. The CNFs might be new generation fertilizers that enhance growth of the plants and defense regulator, also. However, direct application of the carbon-based nanomaterials still remains a concern. To overcome such issues, carbon-based nanomaterials are encapsulated with polymeric composite for agricultural delivery system. Kumar et al. synthesized bi-metallic (Cu/Zn) nanoparticle-dispersed CNFs encapsulated with PVA-starch composite to produce polymer-bi-metal-carbon (PBMC) composite [33]. The produced PBMC polymeric composite is effectively use as a fertilizer that enhances the growth of the plants. The releases of micronutrient (Cu/Zn) from CNFs, as well as polymeric composite in a controlled manner. The study also suggested that the release of micronutrients from PBMC is relatively slow in comparison with CNFs due to encapsulation of polymers. Moreover, CNFs efficiently translocated with the plants through root to shoot to leaves. The produced biodegradable PBMC-based formulation carrying Cu/Zn-CNFs (micronutrients) unwraps newer approach on the application of nanomaterials in agricultures. **Figure 2** shows the

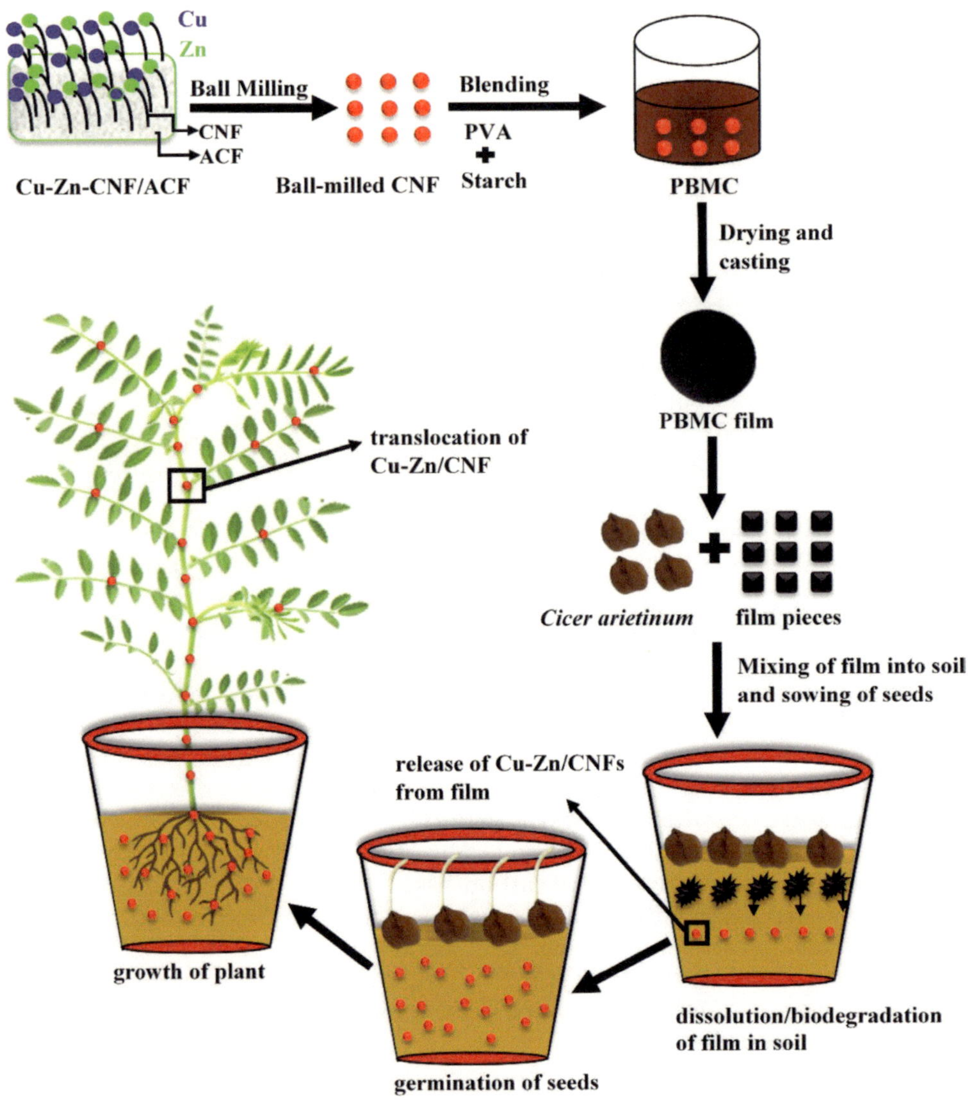

Figure 2.
A schematic representation of synthesis of Cu/Zn-CNF-dispersed polymeric composite and its agricultural application. Reprinted with permission (Kumar et al.), copyright © 2018, Springer Science Business Media, LLC, part of Springer Nature [33].

schematic representation of synthesis of Cu/Zn-CNF-dispersed polymeric composite and its agricultural application.

4. Interaction of polymeric nanocomposite with plants

Interaction of polymeric nanocomposite with plants (accumulation, uptake, and translocation), depends on various factors such as shape, size, surface charge, stability, chemical nature, functional group, and species of the plants. The cell-wall of the plants is one of the major sites of interaction with nanomaterials/other micronutrients. The cell-wall does not permit any foreign particles including nanomaterials/other micronutrients because it acts as a physical barrier. The plant cell-wall contains phosphate, hydroxyl, carboxylate, sulfhydryl, and imidazole groups that produce complex biomolecules, thereby selective translocation and uptake. There are two main properties that affect the uptake and translocation of nanomaterials/other micronutrients: (1) surface charge and (2) size. The surface charge of the nanomaterials/other micronutrients is one of the important parameters. The negatively charged nanomaterials/other micronutrients might favor translocation and uptake within the plants due to negatively charged plant cell-wall. The negatively charged nanomaterials/other micronutrients and plants do not attract each other, thereby easily uptake and translocation of the materials. On the other hand, positively charged nanomaterials/other micronutrients and negatively charged plant cell-wall attract each other, thereby accumulating on the root surface. The metal nanoparticles are positively charged, thereby having high accumulation and less translocation ability. Moreover, these metal nanoparticles also show phytotoxicity at higher concentration due to accumulation [17, 80–83].

The size of the nanomaterials/other micronutrients is one of the important factors for uptake and translocation. The smaller size (20–200 nm) favors the uptake and translocation within the plants. Moreover, carbon-based nanomaterials like CNTs and CNFs ~500 nm or less easily translocate within the plants due to their movement across the epidermis to cortex to vascular bundle. The nanomaterials are translocated to root to shoot to leaves through cell-wall network and plasmodesmata. The capillary action and osmotic forces are also one of the driving forces of translocation of nanomaterials within the plants. Additionally, the types of nanomaterials and chemical composition also affect the uptake and translocation within the plants. The functionalization and coating of nanomaterials alter the adsorption and accumulation ability within the plants. Some of the nanomaterials might accumulate at Casparian strip, whereas another translocate with symplastic routes towards shoot and root [84].

Recently, carbon-based nanomaterials like CNTs and CNFs acted as carriers for genes/micronutrients/biomolecules within the cells. Various are studies performed to understand the exact mechanism behind the nanomaterial uptake and translocation [81]. The larger sized nanomaterials are unable to penetrate cell-walls; however, a study on *Arabidopsis thaliana* leaf suggested the creation of endocytosis-like structure in plasma membrane [85]. Liu et al. suggested that water-soluble SW-CNTs with ~500 nm (length) were exposed on *Nicotiana tabacum.* The water-soluble SW-CNTs are able to penetrate through rigid and integral cell wall [86].

In general, several factors including surface charge, size, chemical nature, and surface coating influence the uptake and translocation ability within the plants [87]. Moreover, functionalization of nanomaterials with chemical/polymer might change the properties of materials, thereby easily translocating within the plants [88, 89].

5. Polymeric nanomaterial improved genetic engineering

Genetic engineering of the plant system is basically efforts of environmental sustainability, synthesis of product, and engineering of agricultural crops; therefore, advancement of genetic engineering is essential for growing population. The gene editing includes various techniques to use for accurately modifying the genome sequence. The emergence of gene editing is an exciting approach especially for agriculture scientist because of the simple process and accuracy that are able to develop improved variety of crops (addition of valuable traits and deletion of antagonistic traits). With the help of genome editing/genetic engineering, researchers continue to focus on the improvement in the yield of the crops with adverse conditions such as changes in climate. Usually, the cell-wall of the plants represents as a physical barrier; therefore, delivery of biomolecules/genes is difficult compared with animal system [90, 91]. Usually, two modes of transformation of genes exist in plants system: (1) cargo delivery that depends on the delivery techniques and (2) regeneration by using transformed plants that depends on the tissues, optimization of the protocols, and complicated hormone mixtures. However, the existing technologies have a lot of limitations such as less transformation, high toxicity, and DNA integration into host genome. The grand challenges of genes/biomolecules cargo delivery within the plants system due to the presence of rigid and multi-layered plant-cell wall, thereby slower transformation of genes/biomolecules within the plants. To overcome such issues, two approaches have been developed and used for transformation of genes/biomolecules within the plants: (1) *Agrobacterium*-mediated delivery system and (2) biolistic particle delivery (DNA bombardment). However, these strategies also have various drawbacks/limitations such as species dependence (changing the species changed the transformation efficiency), required regeneration from tissues, thereby time consuming and less efficiency, and *Agrobacterium*-mediated genes/biomolecule transformation might introduce foreign genetic materials. The *Agrobacterium*-mediated genes/biomolecules might cause disruption of genes/poor/unstable gene expression due to the random DNA integration. The DNA integration might be prevented by using nonintegrated viruses or plasmid deficient in transfer DNA insertion [90, 92]. Therefore, these two strategies are more preferred tools in comparison with other conventional methods. In this context, nanotechnology might be an alternative tool to resolve such issues associated with the existing delivery system.

Various nanomaterial-based plant delivery systems focus on the synthesis of nanomaterials, agrochemical delivery system, micronutrient delivery system, translocation of nanomaterials that augmented the growth of plants by using metal-based nanoparticles, CNTs, CNFs, quantum dots, graphene and its derivatives, and fullerenes. On the other hand, some nanomaterials exhibited phytotoxicity due to the oxidative stress and vascular blockage, damaging the structural DNA. Recently, Demirer et al. [90] developed nanomaterial-mediated biomolecule delivery system for gene expression and silencing of the plant system. For this, grafting of DNA on covalently functionalized pristine SW-CNTs and MW-CNTs was done to produce effective DNA delivery with strong expression of protein in mature *Eruca sativa* (arugula) leaves. The DNA is delivered in plant nucleus with the CNTs and also silencing of functional gene, separately. The grafting of DNA is done on CNTs due to the π-π stacking; the SDS is replaced by adsorption DNA by using the dialysis process. The produced DNA-CNT-based delivery system is comparable to *Agrobacterium*-mediated delivery system. The study also suggested that the produced CNT-based delivery system efficiently expresses protein in arugula protoplasts (cell-wall free) with the transformation (85%) efficiency.

Zhao et al. [93] developed nanoparticle-mediated genetic transformation. For this, they formed the complex of DNA-nanoparticles and delivered into the pollen grains by using magnetic force. The produce approaches to be moderate with insignificant toxicity, genetically stable and transformed plants. These studies suggested that nanomaterial-based delivery system plays a significant role in the advancement in the genetic engineering of the plant system.

In general, genetic engineering of the plant system is more complicated compared with animal system. The approach of the genes/biomolecule transformation within the plants still remains a concern due to the multi-layer and rigid cell-wall. There is lack of effective delivery of the diverse genes/biomolecules within the plant system without damaging the tissues. The nanotechnology might be an alternative tool in the advancement of the genetic engineering in plant systems that resolve such delivery challenge of genes/biomolecules, thereby increasing the utility of genetic engineering.

6. Conclusion and future prospects

Polymeric nanocomposites own distinct features of biodegradability and biocompatibility, which makes it an ideal material to be used in crop protection and micronutrient delivery in the agriculture field. The reactive nanomaterials have been used in various applications because of their functional groups and characteristics; therefore, ENMs might have the potential ability to be used in different applications including agriculture. Moreover, encapsulation of polymers with different nanomaterials like metal/metal-oxide and carbon-based nanomaterials enhanced the controlled release behaviors, biocompatibility, and simple use. Therefore, they are efficiently used in various applications mainly in agriculture. Additionally, uptake, accumulation, and translocation ability of the nanomaterials mainly depend on surface charges, size, and chemical nature of the materials. On the one hand, polymeric coating of nanomaterials might change the functionality and surface charge; therefore, polymeric composite might efficiently translocate within the plants. With regard to advancement in the genetic engineering, nanomaterials might be alternative tools that efficiently delivered genes/biomolecules. Therefore, polymeric nanocomposite enhances the utility of genetic engineering in plant system. As discussed in the text, CNFs is the next generation fertilizer that can easily deliver micronutrients and biomolecules within the plant. However, transformation of these research into field, some issues must be discuss or detailed studies required; (1) cost of the nanofertilizers, (2) safety concern like health/environmental toxicity, and (3) easy applications. We need to do more research in such agricultural areas for easy applicability in the field.

Acknowledgements

The authors acknowledge support from Chulalongkorn University through Chulalongkorn Academic Advancement into its Second Century Project (Small medical Device) and support from NPDF, SERB, Department of Science and Technology, New Delhi, India in the form of a research grant (PDF/2016/003602).

Author details

Mohammad Ashfaq[1,2,3]*, Neetu Talreja[4], Divya Chuahan[5] and Werayut Srituravanich[2,3]*

1 School of Life Science, BS Abdur Rahman Crescent Institute of Science and Technology, Chennai, India

2 Department of Mechanical Engineering, Faculty of Engineering, Chulalongkorn University, Bangkok, Thailand

3 Biomedical Engineering Research Center, Faculty of Engineering, Chulalongkorn University, Bangkok, Thailand

4 Department of Bio-Nanotechnology, Gachon University, South Korea

5 Department of Chemical and Biomedical Engineering, University of South Florida, Florida, USA

*Address all correspondence to: mohdashfaqbiotech@gmail.com and werayut.s@chula.ac.th

References

[1] Fraceto LF et al. Nanotechnology in agriculture: Which innovation potential does it have? Frontiers in Environmental Science. 2016;**4**(20):1-5

[2] Parisi C, Vigani M, Rodríguez-Cerezo E. Agricultural nano-technologies: What are the current possibilities? Nano Today. 2015;**10**(2):124-127

[3] Khot LR et al. Applications of nanomaterials in agricultural production and crop protection: A review. Crop Protection. 2012;**35**:64-70

[4] de Oliveira JL et al. Application of nanotechnology for the encapsulation of botanical insecticides for sustainable agriculture: Prospects and promises. Biotechnology Advances. 2014;**32**(8):1550-1561

[5] Mortensen DA et al. Agriculture in 2050: Recalibrating targets for sustainable intensification. Bioscience. 2017;**67**(4):386-391

[6] Tilman D et al. Global food demand and the sustainable intensification of agriculture. Proceedings of the National Academy of Sciences. 2011;**108**(50):20260-20264

[7] Prasad R, Bhattacharyya A, Nguyen QD. Nanotechnology in sustainable agriculture: Recent developments, challenges, and Perspectives. Frontiers in Microbiology. 2017;**8**:1014-1014

[8] Nair R. Grand challenges in agroecology and land use systems. Frontiers in Environmental Science. 2014;**2**(1):1-4

[9] Aktar MW, Sengupta D, Chowdhury A. Impact of pesticides use in agriculture: Their benefits and hazards. Interdisciplinary Toxicology. 2009;**2**(1):1-12

[10] Zaytseva O, Neumann G. Carbon nanomaterials: Production, impact on plant development, agricultural and environmental applications. Chemical and Biological Technologies in Agriculture. 2016;**3**(1):17

[11] Yuan Z et al. Novel impacts of functionalized multi-walled carbon nanotubes in plants: Promotion of nodulation and nitrogenase activity in the rhizobium-legume system. Nanoscale. 2017;**9**(28):9921-9937

[12] Chichiricco G, Poma A. Penetration and toxicity of nanomaterials in higher plants. Nanomaterials (Basel). 2015;**5**(2):851-873

[13] Lin S et al. Uptake, translocation, and transmission of carbon nanomaterials in rice plants. Small. 2009;**5**(10):1128-1132

[14] Mukherjee A et al. Carbon nanomaterials in agriculture: A critical review. Frontiers in Plant Science. 2016;**7**:172-172

[15] Ghosh M et al. MWCNT uptake in *Allium cepa* root cells induces cytotoxic and genotoxic responses and results in DNA hyper-methylation. Mutation Research. 2015;**774**:49-58

[16] Chen H. Metal based nanoparticles in agricultural system: Behavior, transport, and interaction with plants. Chemical Speciation & Bioavailability. 2018;**30**(1):123-134

[17] Rastogi A et al. Impact of metal and metal oxide nanoparticles on plant: A critical review. Frontiers in Chemistry. 2017;**5**(78):1-16

[18] Gogos A, Knauer K, Bucheli TD. Nanomaterials in plant protection and fertilization: Current state, foreseen applications, and research priorities. Journal of

Agricultural and Food Chemistry. 2012;**60**(39):9781-9792

[19] Rui M et al. Iron oxide nanoparticles as a potential Iron fertilizer for peanut (*Arachis hypogaea*). Frontiers in Plant Science. 2016;7:815-815

[20] Burklew CE et al. Effects of aluminum oxide nanoparticles on the growth, development, and microRNA expression of tobacco (*Nicotiana tabacum*). PLoS One. 2012;**7**(5):e34783-e34783

[21] Belava VN et al. The effect of silver and copper nanoparticles on the wheat-*Pseudocercosporella herpotrichoides* pathosystem. Nanoscale Research Letters. 2017;**12**(1):250-250

[22] Tsi Ndeh N, Maensiri S, Maensiri D. The effect of green synthesized gold nanoparticles on rice germination and roots. Advances in Natural Sciences: Nanoscience and Nanotechnology. 2017;**8**(3):035008

[23] Siddiqi KS, Husen A. Engineered gold nanoparticles and plant adaptation potential. Nanoscale Research Letters. 2016;**11**(1):400-400

[24] Sharma J et al. Role of silver nanoparticles in treatment of plant diseases. In: Patra JK, Das G, Shin H-S, editors. Microbial Biotechnology, Application in Food and Pharmacology. Vol. 2. Singapore: Springer Singapore; 2018. pp. 435-454

[25] Kim SW et al. Antifungal effects of silver nanoparticles (AgNPs) against various plant pathogenic fungi. Mycobiology. 2012;**40**(1):53-58

[26] Korndörfer GH, Lepsch I. Chapter 7 effect of silicon on plant growth and crop yield. In: Datnoff LE, Snyder GH, Korndörfer GH, editors. Studies in Plant Science. Amsterdam: Elsevier; 2001. pp. 133-147

[27] Gangloff WJ et al. Mobility of organic and inorganic zinc fertilizers in soils. Communications in Soil Science and Plant Analysis. 2006;**37**(1-2):199-209

[28] Mortvedt JJ. Crop response to level of water-soluble zinc in granular zinc fertilizers. Fertilizer Research. 1992;**33**(3):249-255

[29] Rajiv P, Rajeshwari S, Venckatesh R. Bio-fabrication of zinc oxide nanoparticles using leaf extract of *Parthenium hysterophorus* L. and its size-dependent antifungal activity against plant fungal pathogens. Spectrochimica Acta Part A: Molecular and Biomolecular Spectroscopy. 2013;**112**:384-387

[30] Wang Y et al. The application of nano-TiO(2) photo semiconductors in agriculture. Nanoscale Research Letters. 2016;**11**(1):529-529

[31] Zhang W et al. Uptake and accumulation of bulk and nanosized cerium oxide particles and ionic cerium by radish (*Raphanus sativus* L.). Journal of Agricultural and Food Chemistry. 2015;**63**(2):382-390

[32] Ashfaq M, Verma N, Khan S. Carbon nanofibers as a micronutrient carrier in plants: Efficient translocation and controlled release of Cu nanoparticles. Environmental Science: Nano. 2017;**4**(1):138-148

[33] Kumar R, Ashfaq M, Verma N. Synthesis of novel PVA–starch formulation-supported Cu–Zn nanoparticle carrying carbon nanofibers as a nanofertilizer: Controlled release of micronutrients. Journal of Materials Science. 2018;**53**(10):7150-7164

[34] Gupta GS, Kumar A, Verma N. Bacterial homoserine lactones as a nanocomposite fertilizer and defense

regulator for chickpeas. Nanotechnology in Environmental Science. 2019;**6**: 1246-1258

[35] Omar RA et al. Impact of nanomaterials in plant systems. In: Prasad R, editor. Plant Nanobionics, Advances in the Understanding of Nanomaterials Research and Applications. Vol. 1. Cham: Springer International Publishing; 2019. pp. 117-140

[36] Jarosiewicz A, Tomaszewska M. Controlled-release NPK fertilizer encapsulated by polymeric membranes. Journal of Agricultural and Food Chemistry. 2003;**51**(2):413-417

[37] Chen BZ et al. In vitro and in vivo assessment of polymer microneedles for controlled transdermal drug delivery. Journal of Drug Targeting. 2018;**26**(8):720-729

[38] Zhang JN et al. Development of a BDDE-crosslinked hyaluronic acid based microneedles patch as a dermal filler for anti-ageing treatment. Journal of Industrial and Engineering Chemistry. 2018;**65**:363-369

[39] Chen BZ et al. Self-implanted tiny needles as alternative to traditional parenteral administrations for controlled transdermal drug delivery. International Journal of Pharmaceutics. 2019;**556**:338-348

[40] Oliveira HC et al. Nanoencapsulation enhances the post-emergence herbicidal activity of atrazine against mustard plants. PLoS One. 2015;**10**(7):e0132971

[41] Roy I, Gupta MN. Smart polymeric materials: Emerging biochemical applications. Chemistry & Biology. 2003;**10**(12):1161-1171

[42] Priya James H et al. Smart polymers for the controlled delivery of drugs: A concise overview. Acta Pharmaceutica Sinica B. 2014;**4**(2):120-127

[43] Talreja N, Kumar D. Engineered nanoparticles' toxicity: Environmental aspects. In: Hussain CM, Mishra AK, editors. Nanotechnology in Environmental Science. 2018. DOI: 10.1002/9783527808854.ch23

[44] Mazzaglia A, Fortunati E, Kenny JM, Torre L, Balestra GM. Nanomaterials in plant protection. In: Axelos MA, Van de Voorde MH, editors. Nanotechnology in Agriculture and Food Science. 2017. DOI: 10.1002/9783527697724.ch7

[45] Hong J, Peralta-Videa JR, Gardea-Torresdey JL. Nanomaterials in agricultural production: Benefits and possible threats? In: Sustainable Nanotechnology and the Environment: Advances and Achievements. American Chemical Society (ACS); 2013. pp. 73-90

[46] Kocak G, Tuncer C, Bütün V. pH-responsive polymers. Polymer Chemistry. 2017;**8**(1):144-176

[47] Kim Y-J, Matsunaga YT. Thermo-responsive polymers and their application as smart biomaterials. Journal of Materials Chemistry B. 2017;**5**(23):4307-4321

[48] Kasirajan S, Ngouajio M. Polyethylene and biodegradable mulches for agricultural applications: A review. Agronomy for Sustainable Development. 2012;**32**(2):501-529

[49] Pascoli M et al. State of the art of polymeric nanoparticles as carrier systems with agricultural applications: A minireview. Energy, Ecology and Environment. 2018;**3**(3):137-148

[50] Van Beilen JB, Poirier Y. Production of renewable polymers from crop plants. The Plant Journal. 2008;**54**(4):684-701

[51] Roy A et al. Controlled pesticide release from biodegradable polymers. Central European Journal of Chemistry. 2014;**12**(4):453-469

[52] Davidson DW, Verma MS, Gu FX. Controlled root targeted delivery of fertilizer using an ionically crosslinked carboxymethyl cellulose hydrogel matrix. Springerplus. 2013;**2**:318-318

[53] Mukherjee R, De S. Preparation, characterization and application of powdered activated carbon-cellulose acetate phthalate mixed matrix membrane for treatment of steel plant effluent. Polymers for Advanced Technologies. 2016;**27**(4):444-459

[54] Wilson HT, Amirkhani M, Taylor AG. Evaluation of gelatin as a biostimulant seed treatment to improve plant performance. Frontiers in Plant Science. 2018;**9**:1006-1006

[55] El Hadrami A et al. Chitosan in plant protection. Marine Drugs. 2010;**8**(4):968-987

[56] de Oliveira JL et al. Geraniol encapsulated in chitosan/gum arabic nanoparticles: A promising system for pest management in sustainable agriculture. Journal of Agricultural and Food Chemistry. 2018;**66**(21):5325-5334

[57] Chang Y-N, Mueller RE, Iannotti EL. Use of low MW polylactic acid and lactide to stimulate growth and yield of soybeans. Plant Growth Regulation. 1996;**19**(3):223-232

[58] Gao H et al. Enhanced plant growth promoting role of mPEG-PLGA-based nanoparticles as an activator protein PeaT1 carrier in wheat (*Triticum aestivum* L.). Journal of Chemical Technology & Biotechnology. 2018;**93**(11):3143-3151

[59] Voinova ON et al. Microbial polymers as a degradable carrier for pesticide delivery. Applied Biochemistry and Microbiology. 2009;**45**(4):384-388

[60] Russo R et al. Alginate/polyvinylalcohol blends for agricultural applications: Structure-properties correlation, mechanical properties and greenhouse effect evaluation. Macromolecular Symposia. 2004;**218**(1):241-250

[61] Paradelo R, Basanta R, Barral MT. Water-holding capacity and plant growth in compost-based substrates modified with polyacrylamide, guar gum or bentonite. Scientia Horticulturae. 2019;**243**:344-349

[62] Lee SS et al. Synergy effects of biochar and polyacrylamide on plants growth and soil erosion control. Environmental Earth Sciences. 2015;**74**(3):2463-2473

[63] Nussinovitch A. Beads and special applications of polymers for agricultural uses. In: Polymer Macro- and Micro-Gel Beads: Fundamentals and Applications. New York, NY: Springer New York; 2010. pp. 231-253

[64] Ponnamma D et al. Synthesis, optimization and applications of ZnO/polymer nanocomposites. Materials Science and Engineering: C. 2019;**98**:1210-1240

[65] Kumar S et al. Nano-based smart pesticide formulations: Emerging opportunities for agriculture. Journal of Controlled Release. 2019;**294**:131-153

[66] Afreen S et al. Carbon-based nanostructured materials for energy and environmental remediation applications. In: Prasad R, Aranda E, editors. Approaches in Bioremediation: The New Era of Environmental Microbiology and Nanobiotechnology. Cham: Springer International Publishing; 2018. pp. 369-392

[67] Ashfaq M et al. Cytotoxic evaluation of the hierarchical web of carbon micronanofibers. Industrial & Engineering Chemistry Research. 2013;**52**(12):4672-4682

[68] Ashfaq M, Khan S, Verma N. Synthesis of PVA-CAP-based biomaterial in situ dispersed with Cu nanoparticles and carbon micro-nanofibers for antibiotic drug delivery applications. Biochemical Engineering Journal. 2014;**90**:79-89

[69] Ashfaq M, Verma N, Khan S. Copper/zinc bimetal nanoparticles-dispersed carbon nanofibers: A novel potential antibiotic material. Materials Science & Engineering. C, Materials for Biological Applications. 2016;**59**:938-947

[70] Ashfaq M, Verma N, Khan S. Novel polymeric composite grafted with metal nanoparticle-dispersed CNFs as a chemiresistive non-destructive fruit sensor material. Materials Chemistry and Physics. 2018;**217**:216-227

[71] Khare P et al. Carbon nanofibers containing metal-doped porous carbon beads for environmental remediation applications. Chemical Engineering Journal. 2013;**229**:72-81

[72] Kumar D, Talreja N. Nickel nanoparticles-doped rhodamine grafted carbon nanofibers as colorimetric probe: Naked eye detection and highly sensitive measurement of aqueous Cr^{3+} and Pb^{2+}. Korean Journal of Chemical Engineering. 2019;**36**(1):126-135

[73] Sankararamakrishnan N, Chauhan D, Dwivedi J. Synthesis of functionalized carbon nanotubes by floating catalytic chemical vapor deposition method and their sorption behavior toward arsenic. Chemical Engineering Journal. 2016;**284**:599-608

[74] Saraswat R et al. Development of novel in situ nickel-doped, phenolic resin-based micro–nano-activated carbon adsorbents for the removal of vitamin B-12. Chemical Engineering Journal. 2012;**197**:250-260

[75] Sharma AK et al. Preparation of novel carbon microfiber/carbon nanofiber-dispersed polyvinyl alcohol-based nanocomposite material for lithium-ion electrolyte battery separator. Materials Science & Engineering. C, Materials for Biological Applications. 2013;**33**(3):1702-1709

[76] Singh S et al. Preparation of surfactant-mediated silver and copper nanoparticles dispersed in hierarchical carbon micro-nanofibers for antibacterial applications. New Biotechnology. 2013;**30**(6):656-665

[77] Talreja N, Kumar D, Verma N. Removal of hexavalent chromium from water using Fe-grown carbon nanofibers containing porous carbon microbeads. Journal of Water Process Engineering. 2014;**3**:34-45

[78] Talreja N, Verma N, Kumar D. Carbon bead-supported ethylene diamine-functionalized carbon nanofibers: An efficient adsorbent for salicylic acid. Clean–Soil, Air, Water. 2016;**44**(11):1461-1470

[79] Bhadauriya P et al. Synthesis of yeast-immobilized and copper nanoparticle-dispersed carbon nanofiber-based diabetic wound dressing material: Simultaneous control of glucose and bacterial infections. ACS Applied Bio Materials. 2018;**1**(2):246-258

[80] Pérez-de-Luque A. Interaction of nanomaterials with plants: What do we need for real applications in agriculture? Frontiers in Environmental Science. 2017;**5**(12):1-7

[81] Rico CM et al. Interaction of nanoparticles with edible plants and their possible implications in the food chain. Journal of Agricultural and Food Chemistry. 2011;**59**(8):3485-3498

[82] Aslani F et al. Effects of engineered nanomaterials on plants growth: An overview. The Scientific World Journal. 2014;**2014**:28

[83] Priyanka N, Venkatachalam P. Biofabricated zinc oxide nanoparticles coated with phycomolecules as novel micronutrient catalysts for stimulating plant growth of cotton. Advances in Natural Sciences: Nanoscience and Nanotechnology. 2016;7(4):045018

[84] Tan W, Peralta-Videa JR, Gardea-Torresdey JL. Interaction of titanium dioxide nanoparticles with soil components and plants: Current knowledge and future research needs–A critical review. Environmental Science: Nano. 2018;**5**(2):257-278

[85] Shen CX et al. Induction of programmed cell death in Arabidopsis and rice by single-wall carbon nanotubes. American Journal of Botany. 2010;**97**(10):1602-1609

[86] Liu Q et al. Carbon nanotubes as molecular transporters for walled plant cells. Nano Letters. 2009;**9**(3):1007-1010

[87] Zhu Z-J et al. Effect of surface charge on the uptake and distribution of gold nanoparticles in four plant species. Environmental Science & Technology. 2012;**46**(22):12391-12398

[88] Kamaly N et al. Degradable controlled-release polymers and polymeric nanoparticles: Mechanisms of controlling drug release. Chemical Reviews. 2016;**116**(4):2602-2663

[89] Mout R et al. Surface functionalization of nanoparticles for nanomedicine. Chemical Society Reviews. 2012;**41**(7):2539-2544

[90] Demirer GS et al. High aspect ratio nanomaterials enable delivery of functional genetic material without DNA integration in mature plants. bioRxiv. 2018;**14**:179549

[91] Abdallah NA, Prakash CS, McHughen AG. Genome editing for crop improvement: Challenges and opportunities. GM Crops & Food. 2015;**6**(4):183-205

[92] Joldersma D, Liu Z. Plant genetics enters the nano age? Journal of Integrative Plant Biology. 2018;**60**(6):446-447

[93] Zhao X et al. Pollen magnetofection for genetic modification with magnetic nanoparticles as gene carriers. Nature Plants. 2017;**3**(12):956-964